Farm Tenancy and the Census in Antebellum Georgia

Farm Tenancy and the Census in Antebellum Georgia

Frederick A. Bode
Donald E. Ginter

THE UNIVERSITY OF GEORGIA PRESS
ATHENS AND LONDON

Paperback edition, 2008
© 1986 by the University of Georgia Press
Athens, Georgia 30602
www.ugapress.org
All rights reserved
Designed by Madelaine Cooke
Set in 10 on 13 Merganthaler Baskerville
Printed digitally in the United States of America

The Library of Congress has cataloged the hardcover
edition of this book as follows:
Library of Congress Cataloging-in-Publication Data

Bode, Frederick A., 1940–
   Farm tenancy and the census in antebellum Georgia /
   Frederick A. Bode, Donald E. Ginter.
Description: xix, 278 p. : maps ; 24 cm.
ISBN 0-8203-0834-X (alk. paper)
   Includes index.
   Bibliography: p. 269-273.
   1. Farm tenancy—Georgia—History—19th century.
2. Georgia—Census, 1860. I. Title. II. Ginter, Donald E.
HD1511.U6 G43   1986
333.33'5 19   85-20946

Paperback ISBN-13: 978-0-8203-3198-0
ISBN-10: 0-8203-3198-8

British Library Cataloging-in-Publication Data available

This book has been published with the help of a grant from the Social Science Federation of Canada, using funds provided by the Social Sciences and Humanities Research Council of Canada.

For Janice and Shirley

Contents

LIST OF APPENDICES	xi
LIST OF TABLES	xiii
LIST OF MAPS	xvii
PREFACE	xix
1 Introduction	1
2 Method	11
3 The Census Variables: A Critique	45
4 The Delineation of Agricultural Regions in Georgia	74
5 The Forms of Tenancy in Antebellum Georgia	90
6 The Spatial Distribution of Tenancy in 1860	114
7 The Transition to 1880	147
8 Conclusion	178
APPENDICES	187
NOTES	247
BIBLIOGRAPHY	269
INDEX	275

Appendices

Appendix A	Estimated Number of Tenants by Convention for Each County (Short Method), 1860	187
Appendix B	Tenancy Rate (%) and Type 3 and Type 5 Estimates by County (Short Method), 1860	197
Appendix C	Long-Method Estimated Tenancy Rates (%) at Four Levels by Militia District, Sixteen Sample Counties, 1860	201
Appendix D	Owsley School Computations of Landowners as Percentage of Total Heads of Household Engaged in Agriculture, by Economic Region, 1850 and 1860	203
Appendix E	Farm Acres and Total Surface Area, by County, 1860	204
Appendix F	Frequency Distribution of Personal Property, Sixteen Sample Counties, 1860	207
Appendix G	Statistics by Militia District, Types 1a, 2a, 3a, and 6a, Sixteen Sample Counties, 1860	223
Appendix H	Statistics by Militia District, Types 4 and 5, Sixteen Sample Counties, 1860	242

Tables

Table 1.1	Minimum and Maximum Tenancy Estimates as Percentages of Free Agricultural Population, Sixteen Sample Counties, 1860	4
Table 2.1	Types b and c as Percentages of Total Schedule IV Entries (Long Method), Sixteen Sample Counties, 1860	16
Table 2.2	Enumerators' Notations on Schedule IV and Dominant Tenancy Convention (Long Method), Sixteen Sample Counties, 1860	22
Table 2.3	Comparison of Results by Short Method and Long Method, Sixteen Sample Counties, 1860	32
Table 2.4	Ratio of Types 4 and 5 to Total Schedule IV Entries (Long Method), Sixteen Sample Counties, 1860	35
Table 2.5	Selected Variables for Five-County Sample in Delta-Loess Region, Mississippi, 1860	35
Table 2.6	Types 3 and 6 as Percentages of Total Schedule IV Entries (Long Method), Sixteen Sample Counties, 1860	38
Table 3.1	Ratio to Total Schedule IV Entries of Farm Laborers and Other Laborers Who Are Heads of Household and Appear on Schedule I Only, Sixteen Sample Counties, 1860	49
Table 3.2	Type 1 Cases Reporting Real Property That Is Greater or Less Than Their Farm Value, Sixteen Sample Counties, 1860	52

Table 3.3	Selected Variables for All Schedule IV Entries Reporting Real Property of $12,000 or More, Harris County, 1860	54
Table 3.4	Comparison of Selected Variables in Harris and Three Neighboring Counties, Aggregate Census, 1860	55
Table 3.5	Improved Acres per Bale of Cotton and as Percentage of Total Farm Acres, Sixteen Sample Counties, 1860, and Baldwin and Neighboring Counties, Aggregate Census, 1860	61
Table 3.6	Comparison with County Aggregates of Selected Variables for Holders of 200 or More Improved Acres, Towns County, 1860	65
Table 3.7	Comparison of Aggregate Personal Property and Estimated Value of Slave Property, Sixteen Sample Counties, 1860	68
Table 3.8	Frequency Distribution for Personal Property of Types 1a, 2a, 3a, 4, 5, and 7a Reporting $0–$499, Sixteen Sample Counties, 1860	71
Table 5.1	Types b and c as Percentage of Total Schedule IV Entries and Ratio of Type 4, Type 5, Farm Laborer Heads of Household, and Other Laborer Heads of Household to Total Schedule IV Entries, Sixteen Sample Counties, 1860	98
Table 5.2	Type 5 as Percentage of Total Types 4 and 5, Sixteen Sample Counties, 1860	100
Table 5.3	Agricultural and Laboring Occupations Designated on Schedule I in Three Counties, 1860	109
Table 6.1	Four Levels of Estimated Tenancy Rates (%), Sixteen Sample Counties, 1860	116
Table 6.2	Comparison of Estimated Tenancy Rates (%) by the Long Method (Levels I and III/IV) and Short Method, Sixteen Sample Counties, 1860	128
Table 6.3	Comparison of Tabulations and Rates (%) Derived by Owsley and by Our Long Method and Short Method, 1860	134

Table 6.4	Comparison of Present Surface Area and Reported Farm Acreage in Selected Militia Districts, 1860	140
Table 6.5	Tenant Farm Size Distribution, 1860	145
Table 7.1	Comparison of Proprietor and Tenant Increases, Sixteen Sample Counties, 1860 and 1880	151
Table 7.2	Tenancy Rates (%) by Race, Nine Counties, 1880	175

Maps

	Georgia Counties, 1860	xx
	Georgia Counties, 1880	xxi
Map 3.1	Farm Acres as Percentage of Total Surface Area	59
Map 4.1	Sixteen Sample Counties Placed within the Economic Regions of Georgia	76
Map 6.1	Tenancy Rates (%), Level I, Sixteen Sample Counties, 1860	118
Map 6.2	Tenancy Rates (%), Level II, Sixteen Sample Counties, 1860	120
Map 6.3	Tenancy Rates (%), Level III, Sixteen Sample Counties, 1860	122
Map 6.4	Tenancy Rates (%), Level IV, Sixteen Sample Counties, 1860	123
Map 6.5	Final Estimates of Maximum Tenancy Rates (Levels III/IV), Sixteen Sample Counties, 1860	127
Map 6.6	Tenancy Rates (%), Short-Method Estimates, 1860	130
Map 7.1	Tenancy (Cash Renters and Share Renters) as Percentage of Farm Operators, 1880	149
Map 7.2	Cash Renters as Percentage of Total Tenants (Statistics), 1880	154
Map 7.3	Cash Renters as Percentage of Total Tenants (Shaded), 1880	155
Map 7.4	Percentage of Change in White Population (Statistics), 1860–1880	157
Map 7.5	Percentage of Change in White Population (Shaded), 1860–1880	159

Map 7.6	Percentage of Change in Black Population (Statistics), 1860–1880	161
Map 7.7	Percentage of Change in Black Population (Shaded), 1860–1880	162
Map 7.8	Slaves as Percentage of Total Population, 1860	164
Map 7.9	Blacks as Percentage of Total Population, 1880	165
Map 7.10	Percentage of Change in Improved Acres, 1860–1880	166
Map 7.11	Percentage of Change in Number of Bales of Cotton, 1859–1879	168
Map 7.12	Pounds of Lint Cotton per Acre under Cotton, 1879	172
Map 7.13	Percentage of Tilled Acres under Cotton, 1879	174

Preface

THOSE WHO SEEK TO REVISE interpretations of the economy and social structure of the antebellum South must do so with trepidation. Giants wage battle in these fields. We have found the giants to be gentle, even when under attack. No one could have been more gracious and supportive of this manuscript than Robert Gallman and Gavin Wright, though they may continue to disagree with some of our interpretations. Stanley Engerman undertook an extraordinarily detailed and careful reading of the manuscript at an early stage and encouraged us to clarify passages where we might have been misinterpreted. Donghyu Yang shared with us the preliminary results of his own studies of North-South wealth inequalities and provided us with important additional data. Useful comments on the manuscript were also received from Dale Swan, Roger Ranson, Richard Sutch, Michael Wayne, and Robert Swierenga. If this book serves to stimulate further debate on the nature of southern rural society, we are content.

Montreal
1985

Georgia Counties, 1860

Georgia Counties, 1880

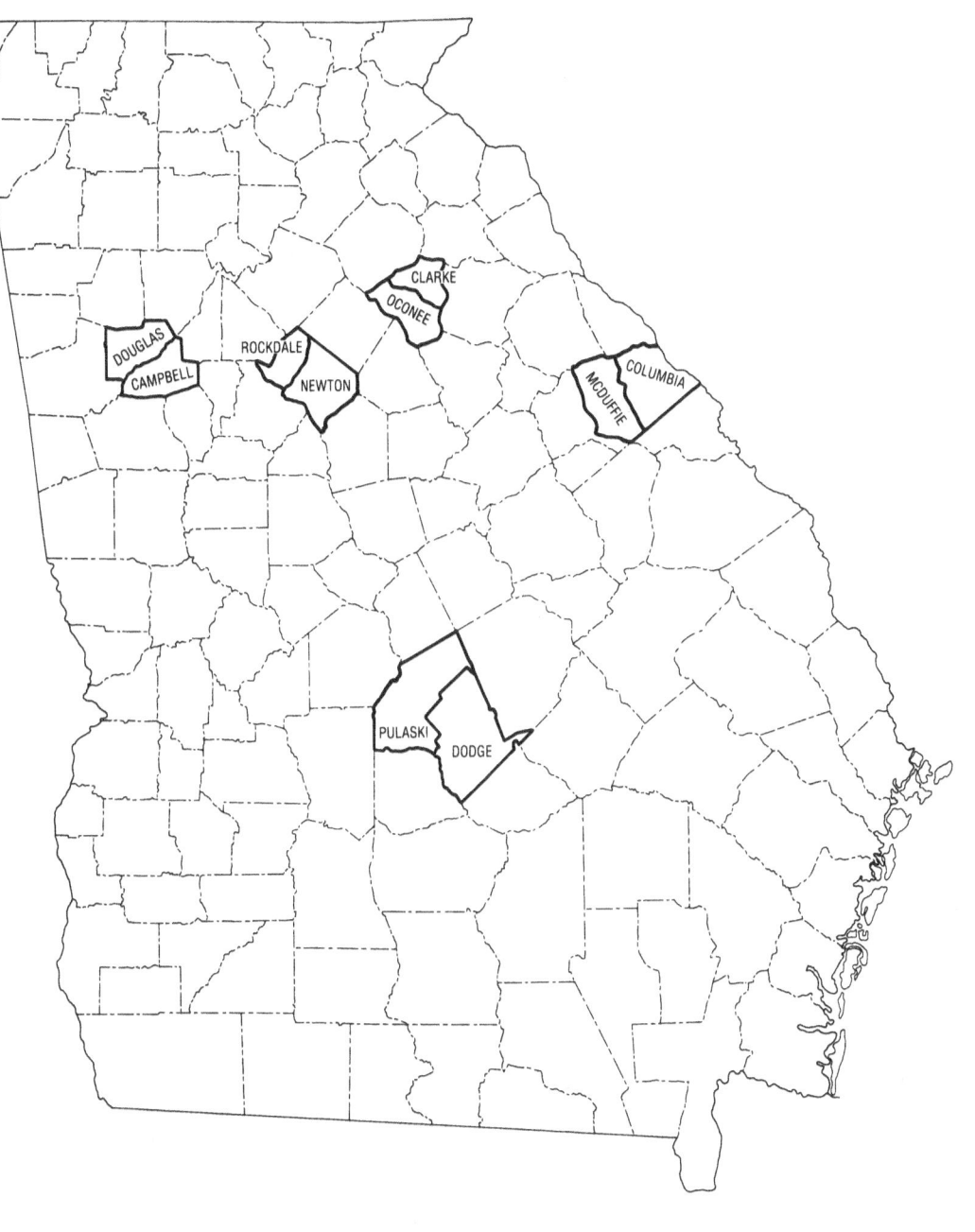

1

Introduction

IN RECENT YEARS ECONOMIC and social historians have shown a vigorous interest in the structure of rural society in the nineteenth-century South. An army of scholars has subjected the antebellum plantation economy to intensive analysis. How slavery affected economic development and how central it was to the very structure of southern society are subjects of lively debate. After years of neglect by historians, the world of the upcountry yeoman farmer is now coming more sharply into focus. Postbellum research has focused importantly on the reorganization of the plantation system, the changing relations among races and social classes, and the spread of tenancy. More broadly, these and other studies have considerably reshaped the old and vexing problem of whether the Civil War and the demise of slavery ushered in fundamental changes in southern society or masked an essential continuity in the region's history. Unfortunately, despite the enormous amount of scholarship devoted directly or indirectly to this theme, historians cannot even agree on how to conceptualize either antebellum or postbellum southern society, much less on how to explain the transition from one to the other.[1]

The "emergence" of tenancy after the Civil War has assumed a central place in the debate. There can be no question that by 1880 most of the freed slaves had become tenants and sharecroppers as family tenancy came to replace the gang system as the principal mode of labor organization and exploitation on plantations. Nevertheless, the timing and mechanism of the transition remain somewhat obscure. Scholars also sharply differ over the nature of tenancy and its impact on southern development. Steven Hahn and Harold D. Woodman insist that postbellum tenancy, and particularly sharecropping, introduced truly capitalist relations on a major scale in the South, with the sharecropper, having no claim to the land or the crop, as proletarian. On the other hand, Jonathan M. Wiener maintains that tenancy perpetuated the "labor-repressive nature of [antebellum] southern production," which was "qualitatively different from the North's classic capitalism." Finally, many economists argue that the various forms of tenancy represented a classic market response to the abolition of slavery and, under postbellum conditions, by and large reflected an equitable distribution of resources, responsibilities, and incomes among the contracting parties.[2]

Despite these fundamental disagreements, there has been a widespread consensus that family tenancy on a significant scale was something new in the South, that it had been at most a marginal feature of antebellum society, and that in the fifteen or so years after the war large numbers of southern whites for the first time lost their independence as owners and came to share with the majority of blacks a dependent status on the land. Roger L. Ransom and Richard Sutch, for instance, maintain that "family tenancy apparently emerged in the Cotton South [after the Civil War] as an almost unprecedented form of labor organization." Wiener asserts that the early postbellum years witnessed "the fall of the white yeoman farmers [in the Alabama hill country] to tenant status." Similarly, Hahn argues that in the upcountry of antebellum Georgia, tenants had been relatively few in number and differed little in their way of life from the landowning yeomanry. One of the few scholars who has taken antebellum tenancy seriously, Joseph D. Reid, Jr., has examined surviving tenancy contracts in one western North Carolina county, but he presents no additional evidence on the scope of tenancy.[3] To some extent scholars have been guided by the pioneering work on antebellum rural society undertaken almost two generations ago by Frank L. Owsley and his students. In his effort to refurbish the historical image of the southern yeoman farmer, Owsley downplayed the significance of tenancy, claiming that 80 to 85 percent of the white agricultural population consisted of landowners. Ransom and Sutch have accepted this estimate, as has Gavin Wright, whose important study of antebellum landholding patterns and the distribution of wealth virtually ignores tenancy.[4] But in the appendix to Owsley's *Plain Folk of the Old South*, and in the separate monographs published by his students, landownership rates appear that are below the ones usually quoted. As we show in appendix D, which summarizes the Owsley school data, landownership rates by their detailed estimates reached 80 percent in only a small minority of regions and were commonly much lower.[5]

Owsley's preconceptions of what life in the Old South must have been like may have led him to misperceive his own data, and later historians, focusing on Owsley's summary estimate, seem to have overlooked the full implications of his findings. Nevertheless, Owsley's work struck us as potentially offering a basis for a substantially revised understanding of the structure of antebellum agriculture. Convinced that a closer look at antebellum tenancy was long overdue, we returned to the manuscript enumerator returns of the 1860 census and, employing a modification of Owsley's method, undertook an analysis of the spatial distribution of tenancy in Georgia. The results of our investigation appear in this book.

Georgia is a good state to examine, since it included within its boundaries numerous regions of greatly differing characteristics: the rice belt on the coast, the sparsely inhabited Pine Barrens and Wiregrass, the cotton lands of the Central Cotton Belt and the lower Piedmont, the hilly country of the upper Piedmont, and the mountains. In some regions small diversified farms were the rule, while in others large staple-producing plantations predominated. Settlement occurred at different times throughout the state. The older counties of the eastern lower Piedmont experienced the first cotton boom, which began in the 1790s. But in the 1850s, while these counties were losing population, the southwest became the state's principal cotton frontier, witnessing a large influx of population, both free and slave. Georgia thus provides ample scope for examining tenancy under a variety of agricultural regimes and demographic conditions.

The principal goal of this study is to establish reliable estimates of tenancy for 1860 in a broadly representative selection of Georgia counties. Although we have some reservations about Owsley's method, and particularly about the form in which he presented his data, we have concluded that in Georgia, at least, he came close to the mark. In part, then, our work confirms the presence of a phenomenon curiously ignored both by its discoverer and by later historians; however, our results also go considerably beyond those of Owsley and his students. Unlike Owsley's, our analysis will extend throughout the entire state of Georgia and will specify tenancy estimates for each of its regions. Our method, while similar to Owsley's in its essential features, will be considerably more refined in its attempt to account systematically for all variations in enumerator technique. Our results are therefore more comprehensive in their coverage and, being more systematically determined, are more empirically reliable.

Since the identification of tenants in the census data proved to be more complicated than we had initially anticipated, we calculated tenancy rates at four levels. The first and fourth levels, which constitute our minimum and maximum estimates for each sample county, are presented in table 1.1. These results, when coupled with rougher estimates for most of the remaining counties of the state, reveal that the incidence of tenancy could vary considerably from county to county. They also show that tenancy rates were often considerably higher, particularly as one approaches the maximum level, than recent scholars have allowed. In general we found that tenancy rates were lowest on the older cotton lands of the lower eastern Piedmont and in the yeomanry-dominated and thinly populated regions of the Wiregrass.

Table 1.1

MINIMUM AND MAXIMUM TENANCY ESTIMATES AS PERCENTAGES OF FREE AGRICULTURAL POPULATION, SIXTEEN SAMPLE COUNTIES, 1860

County	Minimum	Maximum
Baker	17.0	33.3
Baldwin	18.1	26.2
Clay	18.1	26.3
Coffee	27.3	29.6
Echols	20.7	46.2
Emanuel	13.2	17.3
Greene	7.6	17.2
Gwinnett	20.1	32.6
Habersham	37.1	41.6
Harris	19.1	26.3
Houston	10.6	19.6
Thomas	9.2	36.0
Towns	40.6	42.6
Twiggs	31.3	34.9
Wilkes	3.4	14.2
Worth	18.2	29.9

On the newer cotton lands to the west they averaged in the upper 20 percent range, rising to 30 percent and more on the cotton frontier of the southwest and on the cotton fringes of the upper Piedmont, and exceeding 40 percent in the high mountains and on the palmetto flats along the Florida border. White tenancy rates such as these have a striking impact on traditional conceptualizations of southern antebellum society and economy.

The "dual economy" model, as developed by Morton Rothstein and extended by Eugene Genovese and, more recently, by Steven Hahn,[6] may now be seen to fall considerably short of encompassing the fundamental contours of antebellum society and its economic relations. The yeoman sector must be further differentiated between those regions—such as the upper Piedmont, the mountains, and the palmetto flats—where white tenancy was widespread and large landholders were few, and those regions—largely confined to the older lands of the Wiregrass in Georgia—where both tenancy and largeholders were relatively unimportant. The former was a rentier as well as a yeoman economy. The land of the yeoman pure and simple, to the extent such a region existed in antebellum Georgia, was to be found in the latter. A model that presents no more than a dichotomy between regions integrated into a market network and those on the periphery of commercial exchange is clearly inadequate.

If tenancy is to be accepted as a leading feature in some regions of the antebellum southern economy, we may expect a number of other areas of research to be significantly affected. It is too early to judge what the magnitude and direction of those effects might be, but it is not too early to identify their nature. Southern tenancy must now be integrated into our understanding of the westward movement of the region. If tenancies are relatively unavailable on the older lands of Georgia, then younger sons must move west. But their movement westward need not have been so distant as would hitherto have been assumed. The ample availability of tenancies on the lands of western Georgia may be expected to have retarded the more long-distance westward flows of population. Similar regions of tenancy may be expected to have distorted such flows at comparable stages of development in states from Alabama to the Mississippi and beyond.[7] Similarly, there has been a flurry of studies during the last decade investigating the relationship between land availability and family size, the hypothesis being that age of marriage tends to decrease and family size increase when agricultural land is more readily available.[8] Tenancy makes land available, particularly to those, such as younger sons, who have not the capital to purchase. The impact of southern antebellum tenancy for demographic movements and patterns of growth, as well as its obvious implications for upward mobility, pose questions that are now ripe for further investigation.

Our search for tenancy entailed a minute and systematic investigation of landholding variables in the manuscript census. The investigation revealed features of the manuscript entries that have been seriously misinterpreted in the past and that have significant implications for a large portion of the econometric work of the last decade. Much of the difficulty stems from one source: the agricultural returns of the census (schedule IV) often omit acreage and farm value while reporting production. We have been able to conclude that such entries typically are tenants. Entries with missing values in these columns are widespread, and in many counties they comprise 20 to 40 percent of the entries on schedule IV. Briefly, our conclusion is that enumerators in such cases were assigning crops and livestock, which were legally owned by tenants, to the tenants themselves; but they were quietly aggregating the acreages and farm value, which were held but not owned by tenants, under the proprietors. Thus farm outputs were distributed on such returns, but inputs were aggregated; since the schedules do not identify the proprietor of a tenant, no match between inputs and outputs can be made. This is a sobering conclusion for those engaged in arguments regarding economies of scale, for example. But the implications go even further. The Parker-Gallman Sample of Southern Farms 1860, the only machine readable data base drawn from the entire Cotton South in the antebellum period, tended to omit those entries on schedule IV that contained missing values in the acreage and farm value columns. The Sample thus tends to omit tenants, except in that smaller proportion of counties where enumerators included those values for tenants. The Parker-Gallman Sample has played an inestimable role in econometric studies of the American economy since it was drawn in the late 1960s, and it will require some time to draw out the full implications of such findings.[9] Part of that process has already begun. A reevaluation of the Parker-Gallman Sample and the implications of a more fully differentiated southern model are currently evident in comparative studies of the northern and southern economies. Donghyu Yang has suggested revised estimates of

North-South wealth inequalities, based in part on our estimates of southern tenancy entries and their exclusion from the Parker-Gallman Sample.[10] Other revisions will surely follow. While slavery may have been the "peculiar institution" of the South, it is no longer permissible to assume quietly that white farm tenancy was the peculiar institution of the North.

Some studies of southern agriculture, while usually emphasizing the marginality of antebellum tenancy, have nevertheless speculated on its causes and on the forms that it took. In doing so, they have at least suggested hypotheses that might be tested. Some authorities have associated antebellum tenancy with soil exhaustion and poverty. Citing no evidence, Enoch Marvin Banks, writing in 1905, argued that as land became scarce, some people who did not possess the "energy and daring" to start afresh in a distant area became, "in a few cases, tenants on the poorer parts of the large plantations."[11] In 1937 Marjorie Stratford Mendenhall found some fleeting mention of tenancy in antebellum agricultural reports for South Carolina. She concluded, again with little evidence, that the "non-alluvial" planting and farming areas of the state were experiencing "a downward spiral marked by soil erosion and exhaustion, loss of population through emigration, and tenancy."[12] In his classic history of southern agriculture, Lewis Cecil Gray concluded that in the postcolonial South, apart from certain local exceptions, such as the eastern shore of Maryland and Virginia, "tenancy was probably casual, incidental, and transitory." But Gray, unlike Banks and Mendenhall, did not associate tenancy mainly with exhausted soils or population pressure but rather with the clearance of new land. As poorer whites left older regions to "seek their fortunes in areas of cheap land," they were occasionally able to obtain land "free of rent on condition of clearing a certain part of it each year."[13]

There has also been disagreement on the forms of antebellum tenancy. Reid, for example, has claimed that all the varieties of renting and sharecropping that characterized postbellum tenancy had precedents in antebellum practice.[14] Harold D. Woodman, on the other hand, insists that sharecropping was at best marginal before the Civil War. "Sharecropping," he writes, "became the South's new peculiar institution, a unique form of wage labor that grew up on the war-created debris of the old peculiar institution."[15] Finally, Ransom and Sutch have expressed some confidence that at least in the antebellum

Cotton South most tenancy, such as it was, "involved the leasing of plantations or medium-scale farms that were operated by the leaseholder on the plantation system."[16]

Our study will not attempt to deal definitively with all of these issues. We will show that tenancy rates displayed fairly coherent regional patterns within Georgia. These patterns appear to be correlated in complex ways with the importance of staple production, length of settlement, and the availability of alternative supplies of labor. In some counties the enumerators' returns group the data into smaller units called militia districts. Since substantial variation could occur even within a single county, we have chosen three in the cotton belt for more intensive analysis at the district level. At our most speculative, we will consider what light the returns shed on the forms of tenancy in 1860 and suggest they provide considerable evidence for the presence of antebellum sharecropping. Finally, although we have not examined the enumerators' returns for the period after the Civil War, we will offer some observations on the transition to postbellum patterns on the basis of tenancy rates provided by other scholars and by the published aggregate census of 1880.

Another aim of this book is to propose a methodological framework for organizing and interpreting manuscript census data. Therefore, we have not followed the common practice that relegates discussion of data and method to an appendix. We have instead devoted three full chapters and portions of other chapters to these matters. As we began to gather data for this project, we quickly learned, as other scholars had before us, that the census enumerators' returns pose severe difficulties in terms of their intelligibility and the reliability of many variables. Most important, we discovered that enumerators differed among themselves in the way they entered tenants on the agricultural schedule (schedule IV), which purports to list all farm operators producing a specified minimum output, and indeed in whether they entered all, or even any, tenants on that schedule. Most enumerators failed to provide full acreage and farm value data for tenants on schedule IV but employed a limited number of conventions that could exclude some or all of these variables. Some enumerators appear to have recorded some or all tenants only on the population schedule (schedule I). A few enumerators explicitly identified tenants on either schedule I or schedule IV; most did not. Many enumerators also included on schedule IV farm operators and town dwellers who did not meet the minimum production requirements specified in the census instructions.

Our findings indicate that major southern data sources require a more extended and detailed critique than they have yet received. We have found significant and hitherto unidentified structural deficiencies in the 1860 manuscript census. These deficiencies have substantial impact on procedures that have been employed by our predecessors. We would urge that a leading item on the current agenda of quantitative history ought to be a more concerted effort to identify the kinds of problems that exist in the leading data sources and, if not fully resolve them, at least judge their potential seriousness and offer some indication of their magnitude and the direction of their bias. Part of this critical effort should be directed toward the articulation of generally acceptable and standardized classifications of raw data entries. Otherwise the reliability of current findings and their usefulness for comparative analysis may be impaired. We will propose a scheme for classifying agricultural entries from the manuscript census, one that is designed both to minimize and identify error and to achieve some degree of standardization among counties when classifying agricultural populations by landholding category. In the face of enumerator inconsistencies, comparative analysis should proceed only if such standardization is achieved. Even after the application of our scheme, the actual number of tenants in many counties will remain uncertain; hence, we have constructed tenancy estimates at four levels, producing a range between maximum and minimum rates that will allow us and others to make reasoned judgments about the appropriate rate for each county. Different enumerator practices also encouraged us to speculate on the complexity of tenancy arrangements in antebellum Georgia. Despite the fact that enumerators frequently differed among themselves in their procedures, we believe that minimal patterns of intelligibility can be discerned amidst the apparent chaos. For example, we shall present substantial direct and indirect evidence from the census returns that the forms of antebellum tenancy were complex and that the process of white "proletarianization"—sharecropping—predated the Civil War.

Gavin Wright is probably correct in suggesting that "when we encounter studies with such titles as 'How Different Were North and South?' or 'The Postbellum South: Continuity or Discontinuity?' we know in advance that we are not going to get satisfying answers."[17] No fully satisfying answers to such questions will be found between the covers of this volume. This foray into the difficult terrain of southern rural society relies mainly on one source of data, the census; is confined principally to one census year, 1860; and examines a limited number of questions about a neglected topic, antebellum farm tenancy. Obviously a great deal of work remains to be done before anything like a full understanding of the place of tenancy in southern society both before and after the Civil War can be achieved. The results of this study should advance that understanding considerably, however. The traditional conceptualization of antebellum southern social structure can no longer be considered tenable. The fact of antebellum white tenancy does not necessarily alter the basic elements in our understanding of the fundamental role and significance of tenancy in molding the postbellum economy and society. The role and importance of farm tenancy in any society is a continually evolving one and emerges in relation to structural context. But it can no longer be held that tenancy was somehow invented by southerners or emerged suddenly after the Civil War in response to emancipation. The war may have produced many striking elements of discontinuity in the development of the South, but the fundamental elements in its modes of landholding are not among them.

2

Method

THE CENSUS OF 1880 WAS the first to include specific questions on the land tenure of the agricultural population. In that year the enumerator asked each farm operator whether he owned the land in question, rented for a fixed money rental, or rented for a share of the products. The published census reported the number of farmers in each category at the county level. For censuses before 1880, the researcher who wishes to estimate the number of farmers who did not own their land must develop a strategy allowing reasonable inferences only from that data available in the manuscript enumerators' returns.[1]

At first glance the problem seems relatively straightforward. In 1860, the year with which the present study is concerned, the manuscript returns include for every county a schedule of free inhabitants (schedule I) and a schedule for the productions of agriculture (schedule IV) on which the enumerator compiled data on every free person who was a farm operator. Other schedules enumerated the slave population and compiled data on manufacturing, mortality, and social variables.[2] Frequently, but by no means consistently, the county schedules are subdivided into smaller units such as cities, townships, precincts, militia districts, land lottery districts, and post offices. On schedule I each person is listed by name in a sequentially numbered dwelling and separately numbered "family" unit, the reporting of each family purportedly beginning with the head of household, followed by his or her spouse, children in order of age, and all "other inmates, lodgers, and boarders, laborers, domestics, and servants." For each person the following variables are included: age, sex, color, occupation (for males and females over fifteen years of age), value of real estate, value of personal estate, state or (if outside the United States) country of birth, and whether married within the year, attended school within the year, over twenty years of age and cannot read and write, and deaf and dumb, blind, insane, idiotic, pauper, or convict. Schedule IV contains the name of the operator of each farm whose produce was valued at $100 or more, followed by the acres of improved land, the acres of unimproved land, the cash value of the farm, the value of farming implements and machinery, the number of seven kinds of livestock on hand on June 1, the value of all livestock, the quantity of thirty-three kinds of agricultural commodities produced during the year ending June 1, the value of homemade manufactures, and the value of animals slaughtered. With this information it should be possible to locate the name of the farm operator on schedule I, determine if he has real property, and conclude thereby whether or not he owns his own farm. When this process has been completed for each entry on schedule IV, the rate of landownership for the farming population in a county or one of its subdivisions could be calculated. In addition, the other variables on the schedules could provide numerous indicators of the social and economic characteristics of landowners and nonlandowners. But unfortunately the matter is not so simple, as others who have used the manuscript returns have noted.[3]

Facing the researcher are a large number of inconsistencies and ambiguities in the census quite apart from actual errors that an enumerator might have made. Some of these problems arise on schedules prepared by a single enumerator, while others involve inconsistent modes of reporting by different enumerators. It is essential to devise a classification scheme that, if it cannot resolve all difficulties in the enumeration, can at least provide a clear indication of their nature and magnitude. Moreover, such a classification scheme must be consistently applied. The data must, in effect, be standardized. Failure to undertake these procedures could result in invalid inferences from the data and misleading comparisons between geographical units.

The principal findings of this study rely on two distinct research designs. The first design, and the one with which we began our research, comprised an intensive analysis of selected variables from schedules I and IV for sixteen Georgia counties in 1860. For the gathering of the data and the classifying of the farm population in these sample counties, we applied a *long method,* a discussion of which follows.[4] Later in the chapter we will discuss a *short method* for estimating tenancy rates from schedule IV only, a procedure which we subsequently applied to all of the state's 132 counties in 1860 for the purpose of identifying broad regional patterns of tenancy distribution. The chapter concludes with a methodological critique of other studies which have used the census to examine the structure of landholding in the United States before 1880.

The Long Method

For the long method we worked in tandem with schedules I and IV to facilitate the matching of names. This procedure is particularly helpful unless the intention is to create a machine-readable data base that includes all entries on both schedules and to employ a nominal record linkage routine, for matching difficulties regularly arise whenever the enumerator failed to follow the same sequence of names on both schedules. We did not create a machine-readable data base. Our procedures were manual. We kept a running list of all names on schedule IV that could not immediately be located on schedule I and subsequently made another search for them on the latter schedule. Our work sheets contained columns for each variable we intended to examine: occupation, real property, personal property, improved acres, unimproved acres, farm value, and bales of cotton. We also had columns for indicating if the farm operator was not a head of household and if he had no production or possessed only small numbers of livestock.[5] Finally, we devised special codes for any additional relevant information that the enumerator happened to provide, such as whether the farm operator was identified as a manager, agent, or tenant. For each of the sixteen sample counties we included all entries on schedule IV, including those that could not be located on schedule I, and all heads of household on schedule I whose occupation was "farmer" or "planter" but who failed to appear on schedule IV. We recorded the real and personal property holdings, if any, for all such farmer and planter heads of household. We also kept a simple tabulation of the number of heads of household on schedule I who gave their occupation as "laborer," "farm laborer," "farm hand," and the like; however, the significance of this group is substantially vitiated by the fact that large numbers, perhaps most, of those persons assigned laboring occupations did not head their own households but lived with other farm families. Such laborers were not tabulated.

At this stage of our research, in fact, we decided to confine ourselves to heads of household. To have done otherwise would have confronted us with a range of problems that could only be resolved by much additional work with the data. For example, the apparent sons of farmers were usually designated "farmers," sometimes designated "farm laborers," and sometimes given no occupation at all, even if they were over fifteen years of age. The usage of different enumerators could vary widely in this regard. For purposes of comparison, then, it seemed advisable to adhere to categories that could in fact be compared, namely, heads of household. This meant that "farmers" and "laborers" who were living with other families as boarders but who may have been functionally equivalent to those heading a household were lost. The only exceptions to our householder rule were the relatively few schedule IV operators who did not head a household. It should also be noted that in some counties farmers heading a household who failed to appear on schedule IV have probably been undercounted by us because the occasional enumerator either violated his instructions by assigning no occupation to female heads of household or designated them "widows." A full study of the free agricultural labor force will clearly have to take into account not only all those whose occupation related to agriculture and are excluded here but also those members of farm households who were of laboring age but reported no occupation.[6] Such statistics exceed the scope of this study.

Once we had completed the task of entering the variables from the returns on our work sheets, we had to devise a scheme for classifying the various types of records that we encountered on the two schedules. Almost as soon as we began gathering the data, we realized that a neat division between owner-operator and tenant would not be possible. Indeed, as we will show below, the most serious shortcoming of all previous studies of tenancy before 1880 has been the failure to specify appropriate categories and the number of cases in each. A consistent and systematic categorization is necessary to distinguish not only different types of entries made by a single enumerator but also variations between modes of reporting in different counties. For this study we first gave each record a unique numeric code that grouped it into one of the following seven types according to explicit rules that are discussed below.

Type 1: Owner-operator
Type 2: Certain tenant
Type 3: Possible tenant
Type 4: "Farmer" or "planter" heading a household and reporting real property who is entered only on schedule I
Type 5: "Farmer" or "planter" heading a household and reporting no real property who is entered only on schedule I
Type 6: Farm operator on schedule IV who is not entered on schedule I
Type 7: Farm operator on schedule IV who "farms on vacant land" (an enumerator's designation on schedule I found only in Echols County)

We next gave each record of types 1, 2, 3, 6, and 7 (that is, all types found on schedule IV) one of the following alphabetic codes that discriminated, again according to explicit rules, between genuine producers, minor livestock producers, and operators with no production:

Type a: Operator with any arable production (with or without livestock) or, if without arable production, with significant numbers of livestock
Type b: Operator with minor livestock only
Type c: Operator with no arable production and no livestock

Finally, having classified each record according to these typologies, we took note of defective data and other discrepancies that appeared in any particular record. In appendix G we present summary statistics on all our variables, organized into types 1a, 2a, 3a, 6a, and 7a, for each county and, where available, its districts. Appendix H presents statistics for types 4 and 5. The nature and magnitude of all defective and discrepant data are noted. The number of type b and c records for each county may be inferred from table 2.1.

The necessity of distinguishing genuine producers from minor livestock holders and operators with no production (types a, b, and c) arises from the fact that many enumerators failed to follow their instructions. The instructions specified that schedule IV was to include only those farms whose "produce" was worth at least $100 and to exclude "the returns of small lots, owned or worked by persons following mechanical or other pursuits." In a number of counties, how-

Table 2.1

TYPES B AND C AS PERCENTAGES OF TOTAL SCHEDULE IV ENTRIES (LONG METHOD), SIXTEEN SAMPLE COUNTIES, 1860

County	Type b	Type c
Baker	0.0	0.0
Baldwin	0.3	0.0
Clay	0.0	0.0
Coffee	0.6	2.0
Echols	0.0	0.0
Emanuel	0.0	0.0
Greene	0.2	0.2
Gwinnett	0.0	0.0
Habersham	8.8	1.9
Harris	4.7	1.0
Houston	11.7	1.1
Thomas	0.0	0.0
Towns	3.3	1.4
Twiggs	7.2	0.0
Wilkes	0.0	0.0
Worth	0.0	0.0

ever, the enumerators did include "farms," sometimes a large number of them, with either no production at all or perhaps with only one cow or pig, or even only one horse.[7] Failure to allow for such overenumeration of real farms on schedule IV could distort the results for any particular county (if one were measuring, for example, mean farm value) and render comparisons between counties nugatory.[8]

The objective of our research design was twofold: to standardize as much as possible all data across counties, and to do so in a way that restricted the data base to real agricultural producers who must be separately classified from other schedule IV entries. The more difficult problem lay precisely in defining the producer category (type a). Farms with both no arable production and no livestock were easily classified. We simply identified them with a special code, *c*, on our work sheets. The case of small operators, who may nevertheless have fallen under the $100 limit, posed more difficult problems, which we have only partly resolved. Because many enumerators presumably observed such a limit, reconstituting it in errant counties would raise them to a common standard with other counties. If we had included the production of all crops in our data, we could have estimated prices for each crop, which, together with the reported value of animals slaughtered, would have provided a monetary measure that approaches each farm's true output. But we still could not have accounted for the value of livestock that farmers sold on the hoof. In any case, this procedure required time and resources that were not at our disposal and would not be practical for other researchers who, like ourselves, wish to analyze only a limited number of census variables. We therefore sought to define a set of rules identifying in a separate category, in addition to type c, a substantial proportion of operators whose production fell below $100, while at the same time excluding from it genuine producers. Our rules are admittedly imperfect and only partially achieve the kind of standardization between counties that would be optimally desirable.

We accepted as genuine producers all units producing any amount of arable crops and classified them as type a. In a few instances this rule required the acceptance of operators who reported only small amounts of one or two crops, perhaps only some peas and beans, and who almost certainly did not meet the $100 limit. Since such cases were rare, we deemed it reasonable not to try to impose an arbitrary cutoff on arable production.[9] Far more serious and frequent was the presence in some counties of operators who reported only livestock, probably a large number of whom would not have qualified as producers by the census's definition. We devised a rule that allowed us to disaggregate a large number of such operators from schedule IV, giving them the code *b,* minor livestock only. The maximum number of livestock for a type b classification is five in any livestock category, with milch cows, working oxen, and other cattle aggregated into a single category. Thus an operator with three milch cows, two other cattle, and four swine received the minor livestock code. If the same farm had reported one additional cow or ox, or two additional swine, it would not have qualified as a type b. To all farms that exceeded our limits for minor livestock, in addition to those which reported any arable, we applied the code *a.*

Table 2.1 shows the distribution of minor livestock and nonproducing operators (types b and c) as a percentage of all schedule IV entries for each of our sixteen sample counties. It is unlikely that the presence of type c, being relatively rare, would introduce a significant bias into most kinds of statistical analysis, at least in our sample counties. There are several counties, however, where a substantial proportion of operators are type b. We are, however, quite confident that our rule for type b does not impose too high a cutoff on numbers of livestock. The fact that eight of the counties, representing regions with very different kinds of agrarian regimes, have no minor livestock operators, and three others have scarcely any, leads us to conclude that the enumerators in these counties were abiding by their instructions and that type b entries in other counties include very few, if any, genuine producers. Clearly, then, failure to have separately categorized type b operators in such counties as Houston, Habersham, and Twiggs would have seriously affected comparison with other counties.

It remains virtually certain, however, that our cutoff for type b is too low and that as a result in some counties the type a category of genuine producers includes an unknown number of operators who still fall below the census's $100 limit and indeed below any ideal definition of producer. We will ignore operators with insufficient arable production to meet the census minimum, probably a small number in any case, and focus our discussion on the frequently numerous cases of operators who kept only livestock. A consideration of who such operators may have been will shed more light on the adequacy of our type b category in defining a lower limit for genuine producers. We will then suggest an additional strategy for gauging the degree of standardization that one might achieve between counties.

Seddie Cogswell, Jr., in his study of Iowa tenancy, has suggested that farmers with no arable production were mostly recent settlers who had not had time to plant crops that could have been harvested, and hence reported, before the end of the census year, June 1.[10] This hypothesis doubtless has considerable validity. It cannot, however, account for all the large numbers of individuals in some counties who held only *small* amounts of livestock. Our findings show that many of these people were either town dwellers with nonagricultural occupations who could easily have stabled a horse, kept a few pigs, or grazed some cows on someone else's property or on vacant land—precisely those whom the census instructions explicitly required enumerators to exclude—or plantation overseers who kept small numbers of their own livestock. Our type b operators frequently appeared together in large blocks of entries on schedule IV. An examination of schedule I revealed that they were clearly town dwellers. In Houston County, for example, of a total of seventy-six type b operators, thirty-five had nonagricultural occupations, five reported no occupation, twenty-two were overseers, one could not be located on schedule I, and only thirteen were farmers. While we cannot argue our case on any quantitative basis, our experience with the census schedules gives us some assurance that in our counties most individuals who held only livestock, *and who were not actually engaged in farming*, have been correctly classified as type b. A minority of such nonfarming operators have been classified as type a and may not in fact qualify as genuine producers on the basis of the census definition. On the other hand, type a also includes some individuals with nonagricultural occupations, for example, physician or lawyer, who often had extensive agricultural holdings and properly belong in that category.

Houston, as well as other counties in our sample, contained some operators who did not appear to live in towns and who reported their occupation as farmer or planter, but who held only livestock. Some of them fell within our minor livestock category and were classified as type b. We paid insufficient attention to these apparently exceptional cases until we began to examine all of Georgia's counties by the short method. We then found one county, Spalding, whose enumerator provided direct evidence for Cogswell's hypothesis but mainly in the case of *larger* holders of only livestock, those who exceeded our limit for type b. The Spalding enumerator annotated the returns of 23 out of a total of 461 operators as having "just commenced farming this year." With one exception these 23 operators reported no arable production. Seven of these first-year farmers qualified as a type b or c. Spalding's schedule IV return, however, did not contain the nonfarming type b of the sort found, for example, in Houston County. These findings introduced another dimension to the problem of standardizing the data among counties, a dimension for which our type b category may not sufficiently account; for, in addition to enumerators who appeared to respect the $100 minimum and enumerators who included all holders, whether or not they were farmers, of any amount of livestock, there seem to have been others, like the one in Spalding, who entered on schedule IV only persons actually engaged in farming but including first-year farmers with nothing to report or with only livestock.[11] If the Spalding data are any guide, the recording of all first-year farmers should generally produce some proportion of types b and c. It seems unlikely that counties having no types b and c had first-year farmers entered on schedule IV unless, of course, the livestock production of such farmers exceeded the $100 minimum.[12]

The problem raised by our Spalding findings for standardization of the data concerns the statistical significance of the number of first-year farmers whose livestock holdings placed them somewhere between the upper limit for type b and the census minimum for inclusion on schedule IV. If they are statistically significant, then the comparability of counties that report them and counties that do not may be impaired. In Spalding 5 percent of total schedule IV entries were first-year farmers, but of these about 30 percent were types b and c. Therefore, first-year farmers qualifying by our rules as type a constituted 3.5 percent of the county's farm operators. Some of these farmers may yet have fallen below the census qualification of $100. Even if most of them did not qualify for the $100 limit, this is not a large percentage. But Spalding may not be a typical county in this respect, and one can hardly assume that the percentage would not be markedly higher elsewhere.[13]

In sum, our rules discriminating schedule IV entries between types a, b, and c are by no means perfect. Some pseudofarms and possibly most first-year farms have escaped detection, but the consistent application of our rules must have substantially diminished the margin of error and at least partially achieved the goal of standardization. Our experience with Spalding County has, however, suggested to us a way of improving our research design, which unfortunately we could not apply to the data already gathered for this study. We would recommend for future work that in addition to types a, b, and c a fourth category be created, namely, one that would comprise all holders of only livestock who exceed the upper limit for type b. Such a category would undoubtedly include some holders of large herds of livestock who would certainly qualify as genuine producers, otherwise categorized as type a, but it would nevertheless provide the researcher with a basis on which to judge the dimensions of the livestock-only problem and permit the interpretation of results accordingly. While our classification system into types a, b, and c has attempted to deal extensively with the question of standardization, it cannot deal substantively with defining a suitable level for "producer." It is desirable that a working consensus among scholars emerge concerning both of these matters.

The problem of determining the tenure of a farm operator is central to this study. We began with the assumption that genuine farm operators entered on schedule IV but without real property reported on schedule I were tenants, but we were subsequently forced to subject this assumption to a number of qualifications. Inferences from the data had to take into account inconsistencies and ambiguities in the enumeration. The census instructions defined the operator as "the person residing upon or having charge of the farm, whether as owner, agent, or tenant. When owned or managed by more than one person, the name of one only may be entered." The census form itself referred to the "owner, agent, or manager of the farm," clearly confusing some enumerators.[14] One problem, then, is to be able to distinguish between tenants, who by definition possessed no real property, and agents or managers without real property.[15] Enumerators carried out their instructions in a variety of ways. Frequently they indicated on schedule IV, usually in the margin, that the operator was an agent or manager, sometimes giving the name of the owner as well. Less frequently they noted either in the margin or in the acreage columns (and in some Georgia counties in the occupation or real property column of schedule I) that the operator was a tenant. Columns one and two of table 2.2 reveal the extent to which the enumerators in our sample counties followed these practices. In counties where an indication of tenure or management regularly appears, the data pose fewer problems in interpretation. And in counties where only manager or agent is regularly present, actual tenants can also be identified with a high degree of certainty.[16]

The manner of reporting acreage and farm value in the enumeration further supports the proposition that most tenants recorded on schedule IV, as opposed to agents and managers, can be accurately identified. In the process of gathering the data, it became apparent to us that enumerators had followed a limited number of conventions in recording improved acreage, unimproved acreage, and farm value for tenants. Column three of table 2.2 shows that in seven of the sample counties the enumerators did not regularly provide data on any of these three variables for operators we have identified as certain tenants (type 2). Only rarely were all of these variables not recorded for operators engaged in farming who also reported real property on schedule I (type 1a), although they were more frequently not recorded for presumed proprietors with only minor livestock or no production (types 1b and 1c). In two of the sample counties the enumer-

Table 2.2

ENUMERATORS' NOTATIONS ON SCHEDULE IV AND DOMINANT TENANCY CONVENTION (LONG METHOD), SIXTEEN SAMPLE COUNTIES, 1860

County	Tenant or Renter	Agent or Manager	Tenancy Convention[a]
Baker	—	—	2
Baldwin	X	X	2
Clay	X	X[b]	1
Coffee	—	—	2
Echols	X[c]	X[b,e]	3
Emanuel	—	—	4
Greene	—	X	2
Gwinnett	X[b]	—	5[f]
Habersham	X[b]	X[e]	1
Harris	—	X[b]	1
Houston	—	X[d]	1
Thomas	—	X[b]	1
Towns	—	—	1
Twiggs	—	X[d]	1
Wilkes	—	X[d]	5
Worth	X	X[d]	6

[a] The numbers in this column refer to the following tenancy conventions: (1) Improved acres, unimproved acres, and farm value missing; (2) improved acres only; (3) improved acres and farm value only; (4) farm value only; (5) improved acres and unimproved acres only; and (6) improved acres, unimproved acres, and farm value reported.
[b] One to three cases reported in county.
[c] Indicated on schedule I. Also includes some records entered on schedules I and IV and indicated as "farms on vacant land" on schedule I.
[d] Also indicates name of owner in many or all cases.
[e] Indicates name of owner in one case only.
[f] Considerable inconsistency within county in reporting these variables.

ator assigned only improved and unimproved acreage to certain tenants, in one county only improved and farm value, in another only farm value, and in one all three variables.[17] It is unlikely for at least two reasons that these were cases where the enumerator was unable to get the required information and that acreage and farm value, as well as real property on schedule I, are simply missing from the returns. First, since operators with acreage and farm values missing generally reported the value of their personal property, there is no reason to assume that they would not have reported their real property as well. If the enumerator was able to get the one, he would have been just as likely to get the other. Second, and more important, one or more of the three schedule IV variables in question is uniformly missing for farm operators explicitly designated "renters" in Baldwin, Clay, and Echols counties. In Clay all three variables are absent. All but two type 2 records in these three counties were designated "renter." But the practice of excluding values was not applied to explicitly designated agents or managers, to whose farms all enumerators consistently assigned acreage and farm value. Therefore, the absence of real property on schedule I and the presence of an acreage-farm value convention on schedule IV in counties where enumerators did not provide special designations for tenants allow us to classify certain tenants with the considerable assurance that missing values are not simply defective and that managers are being excluded from the tenancy count.

The variety and complexity of tenancy conventions raised additional questions about the identification of tenants in the returns, even suggesting the possibility that different conventions, particularly when used by a single enumerator, were meant to represent distinct forms of tenancy, such as renting and sharecropping.[18] We have not attempted to establish separate categories for each convention found by the long method in our sample counties. We have, however, separately identified as possible tenants (type 3) those farm operators who reported no real property on schedule I but did report improved acreage (though not necessarily unimproved acreage) and farm value on schedule IV in counties where one or both of these values were, by convention, normally not assigned to tenants. These cases may constitute a tenancy convention, but in their mode of reporting acreage and farm value they also resemble schedule IV entries for owner-operators. Such cases could be interpreted as owner-operators for whom the enumerator may have neglected to record real property on schedule I. They may also be managers or agents whom he may have failed to identify as such.[19] While we believe that most of these farm operators were in fact tenants, we established the type 3 category in order to reduce the possibility of an upward bias in our rates for certain tenants.

We classified as certain tenants (type 2) all other farm operators on schedule IV who reported no real property on schedule I. In counties or districts where the dominant tenancy convention was to assign no acreage or farm value to tenants, type 2 includes a very small number of propertyless farm operators who nevertheless reported *either* acreage (improved or unimproved) *or* farm value. Type 2, therefore, consists of all presumed tenants in these counties with the exception of those assigned *both* improved acres and farm value. As we have seen, those that were assigned both are classified as type 3. Notes to the tables in appendix G provide the number of variant type 2a entries, together with the value of all their variables, so that they may be easily disaggregated. In Echols and Worth counties, where both improved acreage and farm value are regularly assigned to all schedule IV entries, the enumerators fortunately identified tenants explicitly. The few operators so identified who also reported real property on schedule I we classified as type 3. We also followed this practice in other counties where operators who possessed real property were designated as tenants; we have noted all such type 3a cases on the tables in appendix G. However, in these same counties we classified as certain tenants (type 2) the two operators who were not overtly designated as tenants but who met all other relevant criteria.

The remaining categories presented fewer problems in definition. We placed in the category owner-operator (type 1) all other schedule IV entries that reported real property on schedule I regardless of the mode of recording their acreage and farm value. We also included in this category public institutions, the lunatic asylum in Baldwin County, for example. We reserved a separate category, type 6, for farm operators not located on schedule I, including all farms operated by an agent or manager when the name of the owner, if provided on schedule IV, did not appear as a separate record on schedule I. One exception to the general rule for type 6 was a very few cases of explicitly designated tenants on schedule IV whom we could not locate on schedule I. These we classified as certain tenants (type 2) as long as the recording of their acreage and farm value corresponded to the county's type 2 pattern.

The presence on the census returns of the "farmer without a farm" has puzzled other investigators and in some cases led them to draw quite misleading conclusions about the structure of landownership.[20] We have, therefore, reserved two categories for heads of household on schedule I whose occupation was "farmer" or "planter" but who do not appear on schedule IV.[21] We have classified these individuals as type 4 if they reported real property and as type 5 if they reported none.

To summarize, we applied to each record on our work sheets type numbers one through seven in accordance with the rules defined above. In addition, we gave each record, except for those in categories 4 and 5, a special code, *a, b,* or *c,* depending on whether the unit in question was, according to our definitions, a genuine farm, reported only minor livestock, or had no production or livestock of any kind. These categories and the rules for applying them resolve as unambiguously as possible the difficulties inherent in the census enumeration and achieve some degree of standardization between counties. Our method placed as many restrictions as seemed at all reasonable on the decision to classify a farm operator as a certain tenant. Any bias in our results is clearly against the assumption of certain tenancy.

A final problem, which does not affect the general classification scheme, concerns those instances where the enumerator failed to record acreage or farm value for farms that clearly ought to have had them. We treated such records as defective for the variables in question and provided them with special codes. Personal property can also be defective in two special circumstances noted below. Our rules are as follows:

1. Any owner-operator (type 1) or other operator appearing only on schedule IV (type 6) but not reporting farm value is defective for that variable.
2. Any owner-operator (type 1) or other operator appearing only on schedule IV (type 6) but not reporting improved acreage is defective for that variable *if* arable production is present. The presence of livestock without arable crops, except for farms in the minor-livestock-only category (type b), requires that at least unimproved acreage be reported. Otherwise, unimproved acreage is defective. In this case, unreported improved acreage is not considered defective. Records in all categories are never defective for unimproved acreage so long as improved acreage is reported.
3. Certain tenants (type 2) can have defective variables only if those variables are regularly reported for a particular county or district.[22] If they are, then the criteria outlined in rules 1 and 2 above apply.

26   Farm Tenancy and the Census in Antebellum Georgia

   4. Possible tenants (type 3) cannot be defective for improved acres or farm value since, by definition, they must possess both.
   5. Operators reporting only minor livestock (type b) or no production (type c) are never defective for improved or unimproved acreage.
   6. Public institutions classified as owner-operators (type 1) and explicitly designated tenants not found on schedule I (classified as type 2) are defective for personal property.

We have indicated the number of defectives for every variable in each category on the tables in appendix G and adjusted all relevant computations accordingly. For example, in computing improved acres per bale of cotton for any category, we have subtracted from the denominator the number of bales produced by those farms that are defective for improved acres. We have not attempted to make any judgment about what ought to be an appropriate acreage or value for any particular farm. The occasional miracle-worker who managed to produce 100 bales of cotton, in addition to other crops, on twenty improved acres retains a place in our data. Considering the wide scope for error that exists in the census in any event, we considered it best not to try to second-guess the enumerators at this level of detail.[23]

The Short Method

The existence of various enumerators' conventions in recording acreage and farm value for tenants suggested to us the possibility of devising a short method for estimating the rate of tenancy from schedule IV for all of Georgia's 132 counties. Such a procedure was desirable in order to buttress our conclusions about the regional pattern of tenancy that emerged from our analysis of the sixteen sample counties. In the end it also shed additional light on the tenancy conventions themselves. In brief, we hypothesized that if the acreage and farm value conventions could be established with a reasonable degree of assurance for a county, it ought to be possible to count the number of such cases on schedule IV and thereby obtain a rough estimate of the number of tenants, thus avoiding the necessity of cross-checking each and every record on schedule I.

   We began with schedule IV and noted the number of possible tenancy conventions, which included any deviation from the normal pattern of reporting full values for improved acres, unimproved acres, and farm value, as well as cases where the enumerator noted that the operator was a tenant, renter, or the like. We then tested the reliability of these apparent conventions by cross-checking at least forty, but as many as ninety, entries from schedule IV with their matches on schedule I. We chose blocks of contiguous entries on schedule IV, beginning usually on the first page, in order to facilitate the matching of names between schedules. The blocks chosen and the number of entries checked depended on the spacing and number of apparent conventions, since we wanted to verify several entries of each convention, as well as entries for whom the enumerator recorded full values. We knew from experience with the long method that any tenancy convention on schedule IV would frequently contain a small number of cases for which the enumerator recorded real property on schedule I and which were in fact defective owner-operators (type 1). In the short method, therefore, we were prepared to accept as genuine tenancy conventions those in which a similar minority of cases, or no cases at all, reported real. For all other apparent conventions, in which a substantial proportion of operators reported real, we considered every case to be an owner-operator. In every county for which our method proved viable we were able to count the number of cases in each reliable tenancy convention. We accepted all such cases as apparent tenants and the balance of entries as owner-operators.[24]

At the same time that we verified the tenancy conventions by cross-checking a sample of entries between the schedules, we kept a count of all farm operators with recorded acreage and farm value whom the enumerator did not explicitly identify as tenants, but who reported no real property on schedule I—the equivalent of possible tenants (type 3) in the classification scheme for the long method. We kept a similar count of any heads of household who gave their occupation as farmer or planter on schedule I but who did not appear as farm operators on the schedule IV pages we were checking—the equivalent of type 5. We then calculated the ratio of each type to the total number of schedule IV entries checked. Since, as we will argue later,[25] many type 5 and type 3 operators were probably tenants, it seemed desirable to have at least a crude approximation of their numbers.[26] We classified each county as low, medium, or high with respect to these types 3 and 5, depending on whether the ratios were 5 percent or less, greater than 5 but not greater than 10 percent, and greater than 10 percent. Finally, we took special note of those counties that contained an unusually large number of householders reporting their occupation as farm laborer or in which the enumerator explicitly identified individuals as tenants on schedule I.

The reliability of the short method for comparative analysis clearly required the reduction of any bias resulting from the presence in many counties of operators whose farm output was less than the census minimum of $100. With the same rules that we developed for the long method, we identified all operators with only minor livestock (type b) and no production (type c), both for presumed proprietors and for each tenancy convention. After subtracting the number of types b and c from each category and from total schedule IV entries, we computed the number of all tenants, as well as those represented by each convention, as a percentage of the county's farm operators.

The final step was to classify each county according to our sense of the reliability of the results and also to identify those counties to which, for one reason or another, we could not apply the short method. We divided all counties into the following five classes:

Class 1: Short method applied with apparently reliable results
Class 2: Short method applied with uncertain results
Class 3: Short method not applied; return too unreliable or names too out of phase between schedules to test efficiently
Class 4: Short method not applied; virtually all schedule IV entries receive full acreage values and farm value, while some number of schedule IV entries tested report no real property on schedule I
Class 5: same as class 4, but virtually all schedule IV entries tested report real property on schedule I

We applied the short method in counties in the first two classes but with apparently more reliable results for class 1. There were three reasons for placing a county in class 2. First, we found inconsistencies in the application of one or more apparent conventions whereby more than a very few cases reported real property on schedule I—not always enough to reject the convention out of hand but enough to cast doubt on our results. In some class 2 counties we accepted some conventions as reliable and counted them but rejected and did not count others as too unreliable. Second, in addition to tenancy conventions, we found a medium to high number of full-value schedule IV entries reporting no real property on schedule I (the equivalent of type 3). Third, we noted the presence on schedule I of individuals explicitly designated as tenants whom we could not locate on schedule IV. If we felt that the reasons for placing a county in class 2 biased our tenancy estimate upward or downward, we added respectively a minus (−) or a plus (+) sign to the results for that county. If the biases appeared to cancel each other out, we applied no sign. We placed in class 1 all counties with very few convention inconsistencies, a low number of type 3 entries, and no schedule I tenants not traceable to schedule IV. It must be reemphasized that these classification decisions were based on a test of a small nonrandom minority of entries on both schedules and therefore should not be taken as definitive.

Tenancy estimates by the short method could not be made for counties in the remaining three classes. Among the counties in class 3 were those whose apparent tenancy conventions proved to be too unreliable—where large numbers of schedule IV entries in these conventions reported real property on schedule I. These represented ten of a total of thirty class 3 counties. The majority of counties in this class, twenty, could not be efficiently tested because the names of schedule IV appeared in a radically different order from those on schedule I. To have taken the time to resolve the order would have violated the *raison d'être* for the short method: a fast, efficient way to estimate tenancy rates. Counties in classes 4 and 5 reported full acreage and farm value data for all, or virtually all, schedule IV entries and had no explicit designation of tenants on that schedule. In other words, there were no unique tenancy conventions that could be counted. We nevertheless tested a minimum of forty schedule IV entries for each of these counties. We placed in class 4 those in which some schedule IV entries appeared to be tenants; that is, they reported no real property on schedule I. A few counties, however, included virtually no tenants on schedule IV, at least on the pages we tested. These counties we placed in class 5. It should be noted that most counties in class 5 reported substantial numbers of heads of household who gave their occupation as farmer or farm laborer and whom we could not trace to schedule IV. As we shall argue later, they were probably tenants.

At this point it would be useful to review the kinds of tenancy conventions that the Georgia enumerators employed on schedule IV in

1860. In using the term *convention*, however, we do not mean to imply that enumerators shared a common rule-governed procedure. We are simply suggesting that the finite number of possibilities for enumerating tenants produced a finite number of modes for such reporting. It does not follow that any one mode reflects an identical form of tenancy in every county where that mode is encountered. Appendix A identifies Georgia's 132 counties by class number and indicates the number of cases in each tenancy convention in class 1 and class 2 counties. It also identifies those counties in which we found an explicit indication of tenancy on schedule I.[27] Not only did conventions vary markedly from county to county, but a single enumerator frequently employed more than one convention in the same county, perhaps in order to distinguish different forms of tenancy.[28] It is possible that whenever a convention is represented by a very small number of entries, such entries may simply be defective for one or more variables. It is noteworthy that only two enumerators in the eighty-three class 1 and class 2 counties consistently recorded full acreage and farm value data for tenants. These two enumerators explicitly identified tenants on schedule IV, and in a few other counties a minority of tenants were so identified and also assigned full values. In addition, the fourteen class 4 counties also apparently reported full values for tenants. The most common practice for enumerators was to leave the acreage and farm value columns on schedule IV blank—what we have called *standard convention*—for all or at least some tenants. In seventy-three class 1 and class 2 counties tenants were entered in standard convention, and in thirty-nine of these counties it was the sole convention. In other conventions the enumerator recorded only improved acres, only farm value, only improved and unimproved acres, or only improved acres and farm value—in other words, any possible combination, except only unimproved acreage, in which one or more of the three relevant variables is missing. For all but one of these conventions, enumerators sometimes indicated—though not always consistently, even for a single convention in the same county—that the farm in question was tenanted. Notations such as "renter," "tenant," and "rents land" usually appear in the margin, name column, or acreage-farm value columns of schedule IV, although in a few counties these notations (also including "cropper," and "farms on vacant land") are found in the occupation or real property column of schedule I. Sometimes the same enumerator used more than one of these designations.

To test the validity of the short method, we applied it to the sixteen sample counties analyzed by the long method and compared the results. Two of these counties, Houston and Thomas, were not analyzable by the short method and were placed in class 3. Of the remaining fourteen, only two, Baker and Echols, exhibit striking discrepancies between the two methods with respect to the percentage of tenants who were genuine producers (type 2a) (see table 2.3). We did place Baker in class 2 and give it a minus sign (indicating an upward bias in our tenancy estimate) because we encountered a fair number of entries in the tenancy conventions that reported real property on schedule I. Even so, we would not have anticipated such a large difference between the two methods. In Echols County the short method clearly failed even to place it in class 2. We gave plus signs to Emanuel and Wilkes because of the presence of type 3 entries on schedule IV, but their presence has no effect on a comparison with type 2a as computed by the long method, since the percentage of type 3 by the long method is separately defined. Baldwin has neither a plus nor a minus sign since, by the short method, the number of discrepant convention entries seemed to cancel out the number of type 3 records. All counties, with the exception of Wilkes, report a larger percentage of type 2a operators by the short method than by the long method; in most cases the difference is quite small. These differences can be explained by the fact that almost invariably at least a small number of apparent tenants turn out to be owner-operators with one or more defective variables.[29] Table 2.3 also shows that the low, medium, and high estimates for types 3 and 5 are, with few exceptions, quite good approximations of the true percentages.

On the whole the short method yields sufficiently reliable results to consider it a viable research tool, although of course it cannot be used in states or regions where enumerators may not have followed tenancy conventions on schedule IV. Specifically, the short method will enable researchers to undertake a typological selection of counties for closer analysis on the basis of their regional representativeness and with respect to differences in conventions. The method should also prove useful in delineating larger regional variations, at least as a first approximation. On the other hand, the cases of Baker and Echols

Table 2.3

COMPARISON OF RESULTS BY SHORT METHOD AND LONG METHOD, SIXTEEN SAMPLE COUNTIES, 1860

County	Class Number	Short Method			Long Method		
		Type 2a as Percentage of Entries on Schedule IV	Type 3a	Type 5a	Type 2a as Percentage of Entries on Schedule IV	Type 3a as Percentage of Entries on Schedule IV	Ratio of Type 5 to Total Entries on IV (× 100)
Baker	2	27.7−	Low	High	17.0	3.3	15.1
Baldwin	2	21.0	Medium	Low	18.1	7.4	2.2
Clay	1	19.2	Low	Medium	18.1	1.4	8.5
Coffee	1	29.4	Low	Low	27.3	1.2	2.0
Echols	1	31.4	Low	Low	23.1[b]	2.5	0.0
Emanuel	2	14.3+	Medium	Low	13.2	1.1	4.1
Greene	1	8.5	Low	Medium	7.6	1.5	10.4
Gwinnett	1	22.1	Low	Medium	20.1	0.8	12.8
Habersham	1	37.8	Low	Low	37.1	0.5	5.5
Harris	1	20.4	Low	Medium	19.1	1.4	8.4
Houston	3	—	—	—	9.9	0.7	9.1
Thomas	3	—	—	—	8.9	4.8	44.0
Towns	1	43.2	Low	Low	40.6	1.2	1.4
Twiggs	1	35.4	Low	Low	31.0	0.4	4.9
Wilkes	2	3.4+	Medium	Low	3.4	7.3	5.0
Worth	1	21.0	Low	Low	18.2	3.8	2.1

[a]Low: 0.0 to 5 percent as a ratio of entries of this type to total entries on schedule IV times 100; medium: greater than 5.0 percent but not greater than 10.0 percent; high: greater than 10.0 percent.
[b]Also includes type 7a, "farms on vacant land."

counties should stand as a warning that the results of the short method for *any particular county* could be highly misleading.

Methodological Critique

Other scholars have made important contributions to census methodology, and their labors have yielded fruitful results. No one, however, has dealt comprehensively with the problems raised by different enumerator procedures. These procedures are often difficult to interpret, but leaving them unresolved can lead to misleading results and, especially, unreliable statistical comparisons among counties. In devising the classification scheme for this study and in identifying tenancy conventions, we have sought to translate enumerator inconsistency into quantitative terms wherever possible. We have done so not just to let the data, properly classified, "speak for itself" but to achieve minimum patterns of intelligibility for the data. We believe that we have been able to discern such patterns, and throughout this study we will offer our interpretations of them. Other scholars may disagree with these interpretations. The point is to provide a reliable evidential basis from which meaningful debate can proceed. The following critique is not intended to disparage the work of others. If we focus on what we believe to be their methodological faults rather than on their virtues, it is only to warn of the traps into which we can all so easily fall. More important, we hope to encourage scholars to greater efforts in devising effective and efficient methods for resolving the difficulties posed by the census returns.

The earliest, and still influential, systematic work on landownership in the antebellum South was undertaken in the late 1930s and the 1940s by Frank L. Owsley, his wife, Harriet C. Owsley, and his students at Vanderbilt University, notably Blanche Henry Clark, Harry L. Coles, Jr., and Herbert Weaver. Later scholars have subjected their conceptualization and statistical technique to sharp criticism, but no one has considered the appropriateness of their method for determining the rate of landownership from the manuscript census.[30] We have learned a great deal from the pioneering work of the Owsley school, and our own procedures are essentially a refinement of theirs. Nevertheless, the presentation of their method is ambiguous, and key aspects of it can only be inferred from their results. Also, Weaver employed procedures that were substantially different from those of his mentor in calculating landownership rates. Fortunately these rates can be recalculated from his own data to achieve a rough comparability.

Owsley, in Alabama and Georgia, and Clark, in Tennessee, followed comparable procedures in their calculation of landowning rates from the enumerator returns for 1850 and 1860. Although Owsley did not fully specify his rules, our reworking of his Georgia data shows that he, like Clark, included in his data base both farm operators listed on schedule IV as well as "farmers without farms" who were heads of household and listed only on schedule I (our types 4 and 5).[31] Owsley's landless farmers, then, comprised propertyless farmers on schedule IV (our types 2 and 3) as well as farmers with no real property, found only on schedule I (our type 5). That the inclusion in the data base of types 4 and 5, landowning and landless farmers, respectively, could distort a rate of landownership computed only from schedule IV entries is demonstrated in table 2.4, which shows the ratio of such type 4

Table 2.4

RATIO OF TYPES 4 AND 5 TO TOTAL SCHEDULE IV ENTRIES (LONG METHOD), SIXTEEN SAMPLE COUNTIES, 1860

County	Type 4 Ratio (× 100)	Type 5 Ratio (× 100)
Baker	6.1	15.1
Baldwin	8.0	2.2
Clay	9.2	8.5
Coffee	1.2	2.0
Echols	0.0	0.0
Emanuel	2.6	4.1
Greene	6.1	10.4
Gwinnett	9.0	12.8
Habersham	0.3	5.5
Harris	2.7	8.4
Houston	1.8	9.1
Thomas	22.9	44.0
Towns	0.2	1.4
Twiggs	0.0	4.9
Wilkes	6.0	5.0
Worth	2.4	2.1

Table 2.5

SELECTED VARIABLES FOR FIVE-COUNTY SAMPLE
IN DELTA-LOESS REGION, MISSISSIPPI, 1860

Landowners	Landowners (%)	Landless	Landless (%)	"O" Improved Acres (%)	"Unidentified" Improved Acres (%)
1,685	85.88	277	14.12	14.12	10.55

Source: Herbert Weaver, *Mississippi Farmers, 1850–1860* (Gloucester, Mass., 1968 [first published, 1945]), 36, table 4; 64.

and 5 farmers to total schedule IV entries in the sixteen sample Georgia counties. It is apparent that in some counties the presence of these farmers will have little or no effect on rates of landownership, but in some they will significantly increase the rate, and in others they will lower it.

As we will show in chapter 6, Owsley's and Clark's rates of landlessness are roughly comparable to our level III tenancy estimates, which include type 4 as proprietors and type 5 as tenants. But types 4 and 5 constitute an anomalous group in the census returns. Their presence in strikingly different proportions from county to county (or their absence or virtual absence from some counties) cannot easily be explained by other structural differences among counties. Whatever one finally decides to do about farmers without farms, it is essential that as a first step they be disaggregated from the rest of the population and presented as separate categories. This Owsley and Clark failed to do. The decision whether to treat those in the type 5 category as tenants and those in the type 4 category as proprietors, or to place them in alternative analytical categories, must then be openly argued rather than assumed. Since Owsley and Clark failed to disaggregate types 4 and 5 as separate categories, their results cannot be reliably compared with those of other researchers who excluded them for the computation of tenancy rates and who also failed to indicate their number.[32]

The rates of landownership and landlessness that Weaver computed for Mississippi in 1850 and 1860 are, as presented in his tables, not directly comparable to either Owsley's and Clark's or our own. He followed procedures that diverged sharply from those of Owsley and that appear rather bewildering.[33] If we take as an example the data he compiled for five sample counties in the Delta-Loess region (table 2.5), we can see that Weaver computed a landless rate of 14.12 percent. This figure is equal to the percentage of farmers in his sample reporting "zero" improved acres. On his acreage table, however, Weaver specified another category, those farmers with "unidentified" improved acreage, amounting to 10.55 percent of the Delta-Loess farmers. The question is, which farmers from the enumerators' returns are included in the "zero" category, which corresponds to the landless group, and which farmers are included in the "unidentified" category? Weaver's data base, like Owsley's, consists of all schedule IV entries plus farmers and planters on schedule I whose names were not found on schedule IV (our types 4 and 5). We inferred that the "zero" and "unidentified" acreage categories comprised type 2 tenants entered in a convention on schedule IV and reporting no improved acres (as well as some number of type 1b or 1c entries that in Georgia frequently did not report improved acres) *and* types 4 and 5 farmers. But which type comprised the "zero" and which type the "unidentified" category Weaver failed to explain. In another context, however, Weaver presented tables with data explicitly drawn from schedule IV and also containing the "unidentified" but not the "zero" category. For the Delta-Loess region the number of cases in these tables is 1,685, equal to the number of landowners Weaver presented earlier. Clearly, then, the "unidentified" category consists of schedule IV operators failing to report improved acres and probably consisting mostly of type 2 tenants in standard convention whom Weaver considered landowners. This discovery clarified a statement by Weaver in his methodological discussion that some enumerators "placed on the agricultural schedule every man who called himself a farmer or planter, whether he was a landowner or not, while others listed only landowners." Apparently Weaver assumed that if an enumerator entered some number of type 5 farmers on schedule I, these were landless farmers, and schedule IV included only landowners regardless of whether they reported real property on schedule I. Indeed, he went on to state, "A man listed on Schedule IV as a landowner should appear on Schedule I with the value of real estate owned. If no real estate value were opposite his name on Schedule I, it was obvious that the entry on Schedule IV was erroneous."[34] In first reading these sentences we assumed that Weaver was probably referring to type 3 entries who reported acreage and farm value on schedule IV but no real property on schedule I. But as his other tables imply, he apparently concluded that in counties with some type 5 entries *all* schedule IV entries, whether or not they reported acreage or farm value, were landowners. Any who failed to report real property on schedule I were errors. We can only assume that if a county contained no (or perhaps few) type 5 records, Weaver must have concluded that schedule IV entries with no real property on schedule I were landless farmers. In all of his regions the number of entries with no improved acres equals the number of landless farmers. These farmers must have been type 5 entries in counties that reported them and type 2 entries in counties where type 5 failed to appear. Type 4 entries, whose presence Weaver noted, must have been excluded from his computations altogether.[35]

In sum, Weaver's landless farmers consisted of some mixture of type 5 and type 2 entries, depending on whether type 5 appeared in any particular county. They almost certainly did not include type 3 since this type would not have had either "zero" or "unidentified" improved acreage. Type 4 entries were completely excluded. Despite this confusion of categories it may still be possible to obtain a very rough comparison between Weaver's data on the one hand and Owsley's, Clark's, and our own on the other. If we assume that most of the farmers in the "unidentified" category were type 2 entered in standard convention, then by combining for each of his regions the percentage of cases in that category with the percentage in the "zero" improved category, we should achieve a landless rate more or less comparable to the level III tenancy rate that we will present in chapter 6. Such comparability assumes that the number of type 4 entries in Mississippi was not large enough to affect the denominator significantly and that the number of type 3 entries was also small. In appendix D, where we present the results of the Owsley studies, we have included Weaver's own estimates of landownership but have added our own revised estimates in parentheses.

All the Owsley studies overlook the problem of standardizing schedule IV data by separately classifying farms with only minor livestock or with no production (types b and c). We have shown that their presence could distort comparability among counties.[36] Moreover, these studies provide no information on the number of operators found only on schedule IV (type 6), nor do they define type 3 as a separate category. The data in table 2.6, which shows the proportion of type 6 and type 3 records to total schedule IV entries in the sixteen sample counties, underscore again the kinds of variations that need to be understood if more reliable inferences about the structure of landholding are to be made.

More recent studies, while often more sophisticated in their statis-

Table 2.6

TYPES 3 AND 6 AS PERCENTAGES OF TOTAL SCHEDULE IV ENTRIES
(LONG METHOD), SIXTEEN SAMPLE COUNTIES, 1860

County	Type 3	Type 6
Baker	3.3	16.5
Baldwin	7.4	6.8
Clay	1.4	2.1
Coffee	1.2	0.0
Echols	2.5	0.0
Emanuel	1.1	0.0
Greene	1.4	2.7
Gwinnett	0.8	0.0
Habersham	0.6	0.4
Harris	1.7	2.0
Houston	0.6	0.8
Thomas	4.8	1.5
Towns	1.4	0.5
Twiggs	0.2	0.8
Wilkes	7.3	2.1
Worth	3.8	3.8

tical apparatus, reveal many of the same kinds of methodological problems as those of Owsley and his students.[37] One phenomenon with which these more recent investigators have attempted to come to terms is the "farmer without a farm" (our types 4 and 5). Merle Curti, in his influential book on Trempealeau County, Wisconsin, from 1850 to 1880, recognized that such farmers had to be treated separately from bona fide operators on schedule IV but then decided to aggregate them together with operators who reported no improved acreage, regardless of whether they reported crop outputs. "We considered that a person with no improved acreage was probably not actually farming," Curti concluded.[38] But was that actually the case? How many such farmers in fact reported crop outputs? Were some of them types b and c? If crop outputs were reported, were the operators with no improved acreage defective for that variable, or was the enumerator in Trempealeau following a convention of not reporting improved acreage for tenants? Whatever they were, such operators disappear into an entirely inappropriate aggregation with types 4 and

5. Finally, Curti did not actually estimate a tenancy rate before 1880, although he did suggest that it was "probably" less in earlier years.[39]

Other scholars have attempted to measure the incidence of tenancy in the Midwest before 1880. Allan G. Bogue assumed that the "minimum number" of tenants in several sample townships in Illinois and Iowa were those operators on schedule IV who possessed no real property (our type 2 and perhaps type 3). He arrived at the "maximum number of tenants possible" by adding propertyless farmers located only on schedule I who were heads of household (our type 5). It is not clear what he did with farmers appearing only on schedule I who reported real property (our type 4). If he failed to add them to the total number of entries on schedule IV, his calculation of a "maximum" tenancy rate could be misleading.[40]

Two scholars, Seddie Cogswell, Jr., and Donald L. Winters, have separately attempted to apply Bogue's method to studies of tenancy in several Iowa counties.[41] After linking all entries on schedule IV with their matches on schedule I, Cogswell accepted those without real property as tenants. Winters's method, however, is more roundabout and, despite his assertion to the contrary, departs from that of Bogue and Cogswell. He first drew up a list of householding "farmers" reporting no real property from schedule I. Next, he compared this list with the entries on schedule IV, a procedure that "permits the culling of non tenants," that is, our type 5. He then calculated a tenancy rate, apparently by taking his pared down list of farmers as a proportion of total schedule IV entries. But Winters's equation excludes from the numerator but retains in the denominator an unknown number of schedule IV operators who might not have reported their occupation as farmer on schedule I. In Georgia, at least, a substantial number of operators, including female householders for whom many enumerators recorded no occupation at all or whom they designated "widows," fell into this category. Neither Cogswell nor Winters alluded to the presence of operators on schedule IV who might not have been traceable to schedule I (our type 6). Indeed, Winters's method did not even allow him to identify such cases.

There are other methodological problems in Cogswell's and Winters's work. "Some" enumerators, according to Cogswell, included farms on schedule IV with "no harvest data." These cases, he inferred, were first-year operators with no crop output to report who were entered on the schedule in violation of the census instructions.[42]

Cogswell provided their number, but merely as an example, for only Clinton County but otherwise made no attempt to discriminate them by categories, such as our types b and c. He apparently included them in his data base, thus potentially restricting the comparability of counties. Clinton County was also examined by Winters, who noted the presence there of the same seventy farms with "no crop data." He added to Cogswell's discussion by revealing that twenty-five of these farms also reported no acreage or farm value, and he excluded them from his data base. In the eleven other counties Winters examined for four census years, he apparently found no other farm operators, tenants, or proprietors who failed to report acreage and farm value. It may be, of course, that Iowa enumerators, unlike their Georgia colleagues, did not follow conventions that frequently excluded one or more of these variables for tenants. On the other hand, neither Winters nor Cogswell referred to the occurrence of even defective records, with acreage or farm value missing, apart from the twenty-five records in Clinton County. Finally, neither scholar provided any information on the number of heads of household who gave their occupation as farmer but failed to appear on schedule IV (our types 4 and 5), although both acknowledged their presence.[43] Both argued that such farmers, whatever their status, need not be presented as separate categories because the Census Bureau, when it began to collect tenancy data in 1880, confined itself to operators on the agricultural schedule. This judgment, it seems to us, mistakenly makes a virtue of precedent. The status of "farmers without farms" may indeed never be known with certainty, but their presence on the returns, which in Georgia varied markedly from county to county, must at least qualify *any possible* inference one makes from a comparative analysis of landholding structure either in any particular census year or between 1880 and previous census years.[44]

Some studies of landownership in the South before 1880, like those of the North, also raise methodological problems. Randolph B. Campbell and Richard G. Lowe have analyzed a sample of 5,000 heads of household in eastern Texas in 1850 and 1860, drawn from schedule I and linked to schedule II (the slave schedule) and schedule IV.[45] From this sample they defined a farming population that they divided into two groups: those "with farms," namely, all entries on schedule IV regardless of occupation, and those "without farms," all entries only on schedule I giving their occupation as farmer (our types 4 and 5). On this basis they reached the following conclusion: "Overall, the percentage of landless farmers in Texas, that is, those who held neither improved nor unimproved acreage [in fact, farmers "without farms" as defined above, *not* schedule IV operators failing to report acreage], was remarkably stable at 19.1 percent... in 1850 and 19.2 percent in 1860."[46] "Landless farmers" thus included our type 4 farmers, who possessed real property, while tenants (our type 2 and possibly type 3) were aggregated together with owner-operators merely because both appeared on schedule IV. They presented a frequency distribution of the improved acreage of the "farm population," showing that 26.2 percent and 23.3 percent in 1850 and 1860, respectively, possessed none; but it is not at all clear whether this farm population included only operators on schedule IV or, in addition, our types 4 and 5. In any case, Campbell and Lowe, without having been aware of it, may have encountered a tenancy convention, in which the enumerator failed to record improved acreage.[47]

The dissertation by Frank Jackson Huffman, Jr., on Clarke County, Georgia, between 1850 and 1880 evinces an awareness of many of the difficulties posed by the enumerators' returns, but the absence of a systematic classification scheme makes his study difficult to use. Huffman defined "renters" as operators on schedule IV with no real property reported on schedule I (our type 2 and perhaps type 3). He excluded "some farms" when matches could not be made (our type 6). He noted that in 1850 the enumerator designated twenty-six farms as rented for which he recorded only improved acreage but that there were "other farms" with both improved and unimproved acreage that also showed no real property on schedule I. There is no indication whether enumerators entered farm value for tenants.[48] His presentation of findings further inhibits recalculation, since he provided no way of determining the number of "farmers without farms" (types 4 and 5) who were heads of household. He aggregated them into two occupational categories: landowning farmers, who included both our type 1 and type 4, and landless agricultural workers, who included our types 2, 3, and 5, as well as farm laborers. One could subtract the owner-operators and tenants from these two groups, but one would still be left with all other members of the population reporting agricultural occupations, farmers, and farm laborers, and not just heads of household.[49] Huffman eliminated eighty-four "farms" from his data base in 1860, because these reported only "small amounts of livestock, dairy products, or corn." Except to state that none of these operators gave his occupation as farmer or reported acreage or farm value, Huffman offered no clear criteria to justify exclusion.[50] Some of Huffman's results, if carefully used, may perhaps be compared with our own. Still, the absence of a clear categorization leaves the matter open to question.

A study by Steven Hahn traces the economic, social, and cultural changes that occurred in two counties of the Georgia upper Piedmont, Carroll and Jackson, between 1850 and 1890 with a view toward explaining the conditions that led to the Populist revolt of the 1890s. Downplaying the significance of antebellum tenancy, Hahn claims that in the upcountry, tenants accounted for only 15 to 20 percent of the agricultural population before the Civil War. On the whole Hahn has made a major contribution to our understanding of the white yeomanry, a group largely neglected by historians since the work of the Owsley school. However, his computation of tenancy rates from 1850 census samples for Carroll and Jackson counties suffers from a number of defects.[51]

Hahn confines the agricultural population to household heads with an agricultural occupation, thus ignoring farm operators on schedule IV who claimed a nonagricultural occupation on schedule I. In Carroll, "farmers" without real estate on schedule I who reported *both* crops *and* acreage on schedule IV constituted under 5 percent of the agricultural population (specified by Hahn as farmers, farm laborers, and overseers), while about 70 percent of other propertyless farmers reported crops but no acreage. The latter group, Hahn suggests, enjoyed "informal tenancy arrangements," so that "tenants of some sort" together accounted for about 21 percent of those engaged in agriculture. Hahn's so-called informal tenants were, of course, entered in standard convention. The predominant use of this convention throughout the state, virtually exclusively in some counties, and the fact that a few enumerators identified such entry types specifically as "tenants" argue strongly against merely informal arrangements. In any case, it is not altogether clear what Hahn means by an informal status.[52] In Jackson County Hahn implies that the enumerator did not enter "informal" tenants on schedule IV, since all potential tenants on that schedule reported acreage as well as crops. Hahn goes on to assume that some proportion of Jackson's type 5 farmers must therefore have been the equivalent of Carroll's informal (in fact, standard convention) tenants. On the basis of this assumption he chooses to add 70 percent of Jackson's type 5s to his group of possible tenants. This results in a tenancy rate for Jackson of about 18 percent. The numerator for Carroll's tenancy rate calculation, then, contains only type 2 "farmers," while for Jackson's it consists of type 2 "farmers" and 70 percent of type 5 entries. In both cases, the denominator aggregates the whole number of schedule I household heads reporting *any* agricultural occupation, including farm laborers. This procedure, however, results in misleadingly low rates, since *tenant* farmers (or at least presumed tenant farmers) are not taken as a percentage only of *all* farmers. If, however, we use Hahn's schedule I samples of household heads and calculate the percentage of *all* farmers who reported no real estate, the landless rates in Carroll and Jackson in 1850 jump to 29.8 and 28.9, respectively. The addition of farm laborers to both numerator and denominator causes the rates to climb still further, to 38.0 and 36.2. These alternative rates demonstrate with unmistakable clarity how different assumptions and procedures can bring about enormous variations in the final results. If the higher rates were accepted for Carroll and Jackson counties, Hahn's conclusions regarding movements in tenancy rates prior to the Civil War would have to be sharply reversed.

We wish to emphasize again that this review of the literature is not undertaken in a spirit that seeks to be destructive of our predecessors' work. We view it rather as an effort to engage in constructive debate about the census, the potential of which as a data source depends on a high degree of scholarly consensus about how it ought to be used. We do believe, however, that scholars have not fully come to terms with the ambiguities in the enumerators' returns. Their failure to develop classification schemes that allow others to judge the magnitude of unusual, discrepant, and clearly distinctive types of entries may cast doubt on the meaning or reliability of even the most ingeniously achieved results. A sensitive, clear, and consistently applied method is the key to uncovering patterns of intelligibility in the data and is as well the key to fruitful comparative analysis. Historical scholarship ought to be past the stage of turning out idiosyncratic case studies and ought instead to be developing genuinely comparable data. We are not arguing that our assumptions are necessarily superior to those of our predecessors, but we urge that until a consensus is achieved on the interpretation of the census, it is desirable that results be presented in a fashion that facilitates recalculation.

We will return, in chapter 5, to our own schema for classifying members of the agricultural population. There we will assess the significance of various entry types in an attempt to make sense of inconsistent procedures among enumerators. We will show, for example, how the relative proportions of types b and c and types 4 and 5 among counties may help resolve the status of "farmers without farms." We will also argue that differences in tenancy conventions suggest more complex patterns in tenancy arrangements that go beyond the simple distinction between a type 1 proprietor and a type 2 tenant. In chapter 6 we will construct tenancy estimates at four levels, each of which depends on different assumptions about which entry types may be classified as a tenant. Our suggestions will necessarily be of a provisional nature, sometimes depending only on plausible conjecture, but we are confident they will justify the emphasis we have placed on an adequate classification scheme as a basis for developing working hypotheses about the structure of agricultural society. But before we undertake an elaboration of classification schemes, it will be useful in chapter 3 to examine more carefully the meaning and reliability of the relevant census variables and in chapter 4 to specify the regional characteristics of Georgia agriculture.

3

The Census Variables: A Critique

THE SELECTION OF VARIABLES for this study was dictated both by the limited number of questions we wished to ask and by the time and manpower we had available for the gathering of the data from the census returns. A more comprehensive analysis of the landholding structure of antebellum Georgia would make use not only of virtually all the information contained in the census but also of supplementary data that might be found in county records, newspapers, and private papers. Our study is more limited in its objectives. We are attempting a preliminary analysis of regional variations in the incidence of tenancy in Georgia in 1860 and the relation of tenancy to the cotton economy. We also address in a provisional way some leading hypotheses regarding the economic condition of antebellum tenants, the forms of antebellum tenancy, and the spatial distribution of tenancy after the Civil War. This chapter reviews our principal considerations in choosing particular census variables and offers some observations on their usefulness and reliability.

We restricted ourselves to the following variables drawn from the population schedule (schedule I) and the agricultural schedule (schedule IV) of the manuscript enumerators' returns of 1860 for the sixteen sample Georgia counties examined by the long method:[1]

1. Occupation (I)
2. Value of real estate owned (I)
3. Value of personal estate owned (I)
4. Number of improved farm acres (IV)
5. Number of unimproved farm acres (IV)
6. Cash value of farm (IV)
7. Number of 400-pound bales of ginned cotton (IV)

Our data base consisted of all operators listed on schedule IV, whether or not they could be linked to entries on schedule I, as well as all heads of household on schedule I who gave their occupation as farmer or planter, whether or not they were traceable to schedule IV. The number of potential variables for each record depended on whether it could be found on both schedules. We also made a simple count of "farm laborers" and other "laborers" who were heads of household and who were listed on schedule I only. Our procedure was to transcribe on our work sheets all variables exactly as they appeared on the returns, even in cases where the reported value seemed implausible. We subsequently standardized the spelling and expanded abbreviations of occupations before compiling them on our final tables; otherwise we left the occupational designations largely unaltered.[2]

Occupation

Occupation is an essential variable for a preliminary analysis of tenancy. One could, for example, quite plausibly hypothesize that a disproportionate number of tenants in the antebellum South were principally engaged in nonagricultural pursuits and rented some land only in order to supplement their income. If this hypothesis were true, the significance of tenancy as a step in the southern agricultural ladder would be much diminished.[3] In fact, the great majority of tenants entered on schedule IV, like the great majority of proprietors, in every county examined by the long method reported an agricultural occupation. Indeed, there does not appear to have been any marked difference between the two groups in this regard, although the range of nonagricultural occupations among proprietors was greater.[4] Occupational data were also essential, as we showed in chapter 2, for resolving some of the problems associated with the presence on schedule IV in some counties of operators who reported no production or only small numbers of livestock (types b and c). Finally, the inclusion of "farmers" listed as heads of household on schedule I who could not be traced to schedule IV (types 4 and 5) and of "farm laborers" and "laborers" heading households on schedule I, all of whom were found in strikingly varying proportions between counties, raised questions about the essential meaning of the occupational designations in agriculture and whether they had any systematic relationship to the structure of agricultural tenure and labor.

The census instructions attempted to introduce a measure of consistency into the recording of occupations. In the first place, the enumerator was to "record the occupation of every human being, male and female, (over 15) who has an occupation or means of living," in a manner "so clear as to leave no doubt on the subject." In the second place, the enumerator was instructed to use specific designations for certain kinds of occupations and, by implication, not merely follow his own preferences or put down whatever the respondent told him. In the case of agricultural occupations, for example, the "proprietor of a farm for the time being, who pursues agriculture professionally or practically, is to be recorded as a farmer; the men employed for wages by him are to be termed farm laborers." Finally, for those who had more than one occupation only "the name of the most prominent" was to be recorded.[5]

Despite the efforts of the census office, the problem of sorting out the meaning of occupations is made more difficult by the inconsistent practices of enumerators. Sometimes these inconsistencies resulted from the obvious failure of some enumerators to follow their instructions. Many enumerators regularly entered dual occupations, such as farmer and lawyer; this was clearly a violation of the instructions; so, too, was the practice of some enumerators who recorded no occupation for female heads of household, or designated them "widows," even when such persons appeared on schedule IV as the operator of a farm. Moreover, enumerators were inconsistent in applying the instructions to other heads of household who were the proprietors of farms. In addition to the designation "farmer," some enumerators entered "planter," and not always just for large operators for whom the latter title was traditionally reserved in the South. Even a few tenants turned up as planters.[6] It also appears likely that some enumerators violated their instructions by not distinguishing farm laborers from other types of laborers. In the almost exclusively agricultural county of Echols, for example, the ratio of household heads on schedule I designated "laborer" to total schedule IV entries was 43 percent. At the same time Echols reported no householders who were "farm laborers."[7]

Enumerator inconsistencies with respect to occupation also appear to have resulted from ambiguities and omissions in the instructions themselves, which, in the case of agriculture, provided reasonably clear guidelines only for farm proprietors and those working on a farm for wages. How the enumerator was to designate the occupation of a farm tenant or the spouse and children of a farmer was left unstated. We have already commented on the variety of enumerator practices for entering the occupations of the other members of households headed by a farmer, a problem that is not of immediate concern to this study whose data base is confined almost entirely to heads of household.[8] A more important issue has to do with the occupational designation of tenants and other householders, besides proprietors, engaged in agriculture. It is true that enumerators in most Georgia counties typically entered "farmer" in the occupation column on schedule I for tenants who appeared as farm operators on schedule IV, both in the case of certain tenants (type 2) and possible tenants (type 3).[9] But most enumerators in our sample counties also placed a small minority of tenants, whom we have classified as genuine producers (type 2a), in a laboring occupation, usually "farm laborer," but also "farm hand," "day laborer," "laborer," or "hireling."[10] Perhaps some enumerators intended to make real distinctions between different kinds of tenants. But some may have recorded whatever occupation the respondent happened to give them, while others may just have been whimsically inconsistent. It is also possible that some persons may have divided their time between tenancy and a laboring occupation. There are as well at least three counties in Georgia—Madison, Paulding, and Walton—in which the occupational designation for some entries on schedule I indicates tenancy. In Paulding, for example, the enumerator distinguished between such occupations as "tenant," "renter," and "cropper."[11]

The meaning of laboring occupations for heads of household entered only on schedule I is also frequently unclear. Table 3.1 divides such persons in our sample counties between "farm laborers" (or "farm hands") and *apparent* nonagricultural "laborers" (or "day laborers" or "hirelings") and presents them as a ratio of total schedule IV entries.[12] For one thing, it is surprising that seven counties reported no farm laborers who were heads of household and that two of them, Coffee and Towns, reported no laborers of any kind. Greene and Wilkes, otherwise very similar counties located on the old cotton lands of the eastern lower Piedmont, had radically different proportions of farm laborers. Wilkes, however, was the only major cotton-producing county in our sample to report no one designated "overseer." But a substantial number of farm laborers were explicitly designated "manager" on schedule IV and were thus overseers in fact,

Table 3.1

RATIO TO TOTAL SCHEDULE IV ENTRIES OF FARM LABORERS AND OTHER LABORERS WHO ARE HEADS OF HOUSEHOLD AND APPEAR ON SCHEDULE I ONLY, SIXTEEN SAMPLE COUNTIES, 1860

County	Farm Laborer Ratio[a] (× 100)	Other Laborer Ratio[b] (× 100)
Baker	7.5	1.4
Baldwin	1.6	15.1
Clay	3.9	0.4
Coffee	—	—
Echols	—	43.0
Emanuel	2.2	—
Greene	0.7	3.9
Gwinnett	8.7	3.7
Habersham	—	0.9
Harris	—	1.8
Houston	1.1	2.6
Thomas	—	4.5
Towns	—	—
Twiggs	—	3.2
Wilkes	18.3	—
Worth	9.4	4.2

[a] Includes "farm laborer" and "farm hand."
[b] Includes "laborer," "common laborer," "day laborer," "hireling," and "workman."

even if not so called by the enumerator. Some of the apparently irrational patterns exhibited in table 3.1 may be related to the varying proportions in these counties of "farmers" who were heads of household and appeared only on schedule I, especially those without real property (type 5) (see table 2.4). Some type 5 farmers could have been farm laborers working for a daily, weekly, or monthly wage or possibly tenants of some kind. There is also evidence that some enumerators placed tenants in a laboring occupation, sometimes entering them on both schedules but sometimes only on schedule I.[13] In any event, occupational designations as such offer few clues for resolving these issues.

The ambiguity of the census instructions and the inconsistent and sometimes contradictory practices of enumerators make the analysis of occupations from the census extraordinarily complex, even in the agricultural sector. It may generally be possible to distinguish agricultural from nonagricultural occupations, the apparent confusion between farm laborers and laborers in some counties being a notable exception; but in themselves the designations offer no clear-cut guidelines for fully understanding the occupational structure of the agricultural population. The complexity increases as one moves downward from proprietor to tenant to laborer. Moreover, comparisons of occupational frequency distributions, even for heads of household, among counties and between census years, is of doubtful validity to the extent that enumerators employed noncomparable criteria.

Real Property and Farm Value

The presence or absence of the value of real estate on schedule I is the key variable for determining whether a farm operator on schedule IV was a proprietor or tenant.[14] Farm value on schedule IV could be an important variable for comparing differences in the rental values of tenant-operated farms in a single region and between regions. Together with acreage data and personal-property data, farm value could tell us much about the relative amount of resources available to tenants and proprietors. Unfortunately, farm value was regularly recorded for tenants in only three of our sixteen sample counties.[15]

Real-property value and farm value are treated together in this section, since one would expect to find a more or less systematic relationship between them on the census returns. The enumerators were instructed to "insert [on schedule I] the value of real estate owned by each individual enumerated . . . be the estate located where it may." Moreover, he was not to abate the value because of any "lien or encumbrance." The cash value of the farm was to "include the actual cash value of the whole number of acres returned by you as improved and unimproved." Since the instructions for improved and unimproved acreage allowed that such land need not have constituted contiguous parcels, one would expect that a single farm operator would have been entered only once on schedule IV in a single county or at least in a single district. In fact we failed to find a single instance when the same individual was listed more than once on schedule IV within a single county. The implication, then, is reasonably clear that the *total* value of a person's real property, wherever that property may have been located, would have been reported on schedule I only once, in the county of residence. Farm value, on the other hand, would have been reported separately in connection with each set of farm holdings in one or more counties. For example, the total value of real property of a person who owned a farm and resided in Greene County, but who also owned a farm in Houston, would have been reported on schedule I in Greene, the county of residence. But the value of each of his farms would have been reported on schedule IV separately in the two counties, as a type 1 in Greene and as a type 6 in Houston.

We expected that in general the value of an individual's farm would equal the value of his real property. In most cases we found this to be true. Whenever there were differences, real property was usually greater. Some farmers owned property outside their county of residence, which would not have been recorded as part of the value of the farm in that county. It is well known that large planters in particular operated multiple holdings. Frequently they were residents of older counties who invested in land in newer regions. John B. Lamar of Bibb County, for example, operated a number of plantations throughout Georgia and northern Florida.[16] One resident of Baldwin County reported $388,040 of real property in 1860 but operated a farm there worth only $6,800. Baker County, located in the newer cotton lands of southwest Georgia, counted thirty-five absentee proprietors in 1860, who, although they comprised only 16.5 percent of the farm operators in the county, accounted for 45.2 percent of the farm value.[17] Excess real property could also have included land within the county that was not part of a farm—town lots, for example, or other unused property. There were also some cases where real-property value was less than farm value. These could have been farms owned jointly by two or more persons, the census instructions having allowed the enumerator to enter only one name on jointly owned properties.[18] A second possibility is that some farmers rented parts of their farms, with the difference between real-property value and farm value constituting the value of the rented land. One can also not rule out the possibility of enumerator error in cases where real-property value was either more or less than farm value.[19]

There is evidence, however, that some enumerators may have violated their instructions with respect to real-property value and farm value. Table 3.2 shows the percentage of type 1 (owner-operator)

Table 3.2

TYPE 1 CASES REPORTING REAL PROPERTY THAT IS GREATER OR LESS THAN THEIR FARM VALUE, SIXTEEN SAMPLE COUNTIES, 1860

County	Real Property Greater (%)	Real Property Less (%)
Baker	20.3	14.6
Baldwin	29.4	7.6
Clay	6.8	5.0
Coffee	4.2	2.5
Echols	15.6	1.1
Emanuel	90.6	6.5
Greene	20.6	7.2
Gwinnett	12.7	3.1
Habersham	6.2	1.0
Harris	3.8	4.1
Houston	8.7	7.7
Thomas	28.0	8.0
Towns	10.0	6.1
Twiggs	18.6	3.6
Wilkes	20.2	8.5
Worth	7.1	3.3

Note: All type 1 records that are defective for farm value are excluded.

cases for which the value of real property was either greater or less than farm value in our sample counties. The clearest example of erroneous reporting comes from Emanuel County, where real-property value is greater than farm value for more than 90 percent of owner-operators. In many of these cases the difference between the two values is considerable. The most plausible explanation for this large discrepancy is that the Emanuel enumerator, in recording farm value, was entering the value of *improved* acres only. According to our schedule IV data, Emanuel in 1860 had a total of 506,434 farm acres, improved and unimproved, and a total farm value of $239,191, yielding an average value per acre of forty-seven cents. In the same year the comptroller general of Georgia reported 539,278 acres for taxation in the county valued at $632,874, or $1.17 per acre.[20] The total value of *real* property for all schedule IV entries was $625,758 which, if divided by the total farm acreage, results in an average value per acre of

$1.24, an amount quite close to that reported by the comptroller general.[21] If we assume that the Emanuel enumerator was recording only the value of improved acres, then the average value of such acreage in 1860, according to our census data, was $6.17. The average value of *all* improved and unimproved acreage in the state in that year, as calculated from the data in the aggregate census, was $5.89. Since improved acreage constituted only 30.2 percent of the total farm acreage in the state, almost certainly the average value of the state's improved acreage would have been substantially greater than the $6.17 average in Emanuel. Such a result would be consistent with Emanuel's location on the poor agricultural lands of the Wiregrass region.

In at least one county, Harris, there is fairly convincing evidence that the enumerator violated his instructions by systematically equating the value of real property and farm value. In any county, those cases for which the value of the real property of farm operators exceeded their farm value should normally represent persons who held either nonfarm property within the county or property of any kind outside the county. Table 3.2 shows that of our sample counties Harris had the lowest proportion of such cases, 3.8 percent. On the face of it, this figure is surprising, because Harris, a major cotton producing county on the Alabama border in the lower Piedmont, was precisely the sort of place where one ought to have found substantial numbers of farmers with other holdings elsewhere. One would also expect that the persons most likely to have had multiple holdings would have been the wealthiest planters. Yet, as table 3.3 indicates, only one out of the nineteen farm operators in Harris County with real property valued at $12,000 or more reported a farm value that was less than real-property value. One other individual reported more farm value than real-property value. Conversely, the mean value of real property for the twenty-one operators who reported more real-property value than farm value was $4,208.33, with only four of these twenty-one individuals reporting more than $6,000 in real estate. It is clear that proprietors reporting greater real-property value than farm value were predominantly small holders and not the type of farmer one most expects to have had holdings outside the county of residence. The most likely explanation for the presence of these twenty-one cases in the returns, then, is enumerator error.[22]

In equating the value of real property and farm value, the Harris

Table 3.3

SELECTED VARIABLES FOR ALL SCHEDULE IV ENTRIES
REPORTING REAL PROPERTY OF $12,000 OR MORE, HARRIS COUNTY, 1860

Real Property ($)	Improved Acres	Unimproved Acres	Farm Value ($)	Farm Value/ Acre ($)
38,000	2,000	1,600	38,000	10.56
32,200	1,600	4,830	32,200-	5.01
22,654	1,000	2,369	22,654	6.72
22,500	750	600	22,500	16.67
19,905	2,200	1,381	17,105	4.78
18,087	620	526	18,087	15.78
18,080	1,050	1,430	18,080	7.29
18,000	985	1,075	18,000	8.74
17,500	1,000	2,500	17,500	5.00
17,200	315	3,135	17,200	4.98
17,000	1,150	500	17,000	10.30
16,000	2,000	620	16,000	6.11
14,700	1,700	370	15,700	7.58
14,000	750	450	14,000	11.67
12,000	125	425	12,000	21.82
12,000	800	1,500	12,000	5.22
12,000	1,000	1,100	12,000	5.71
12,000	500	2,000	12,000	4.80
12,000	600	600	12,000	10.00

enumerator could have been violating his instructions in one of two ways. First, he might have *undercounted* the true value of real property by entering an individual's holdings only in Harris, the county of residence. Alternatively, he might have *overcounted* farm values (as well, perhaps, as acreage and other agricultural variables on schedule IV) by including in it the value of farm holdings outside the county. The Harris returns could be defective, then, either by the underreporting of real-property value or the overreporting of farm value and possibly other schedule IV variables. Table 3.4 compares selected aggregate census variables for Harris County with those of three bordering counties.[23] The values of all the variables are more consistent with the first alternative—that the enumerator undercounted the value of real property in Harris—than they are with the second alternative—that

Table 3.4

COMPARISON OF SELECTED VARIABLES IN HARRIS AND THREE NEIGHBORING COUNTIES, AGGREGATE CENSUS, 1860

County	Improved Acres as Percentage of Total Farm Acres	Total Farm Acres as Percentage of Total Surface Area	Farms Reporting 500–999 Improved Acres as Percentage of All Farms	Farms Reporting 1,000 or More Improved Acres as Percentage of All Farms	Farm Value per Total Farm Acre ($)	Real Property per Household ($)
Harris	52.9	97.8	9.1	2.5	6.57	1,724.41
Meriwether	53.0	96.2	9.2	2.2	7.92	2,170.13
Talbot	55.0	96.9	14.8	5.2	8.09	1,501.07
Troup	56.3	90.8	9.0	2.6	8.45	2,454.08

he overcounted farm value and other schedule IV variables. If the second alternative were true, Harris ought to have yielded significantly higher values for each of the variables on the table.[24] There is no indication that Harris reported a disproportionate number of large holdings or substantially more farm acres in relation to its area than the three other counties. Farm value per total acre was indeed lowest of all in Harris, and only one other county reported less real-property value per household. The only evidence consistent with the hypothesis that the Harris enumerator reported the farm value of all holdings, whether inside or outside the county, can be found on table 3.3, which shows that the farm value per total acre for some of the wealthiest planters was much greater than the county average of $6.57. But in this case the enumerator would inconsistently have had to report only the farm acreage actually located in Harris. Moreover, in any county, one may expect to find some number of holdings on prime lands reflecting substantially higher acreage values. But a more conclusive judgment about the procedure of this enumerator would require an analysis of the manuscript census returns for the neighboring counties in order to determine the proportion of farm operators whose real-property value exceeded their farm value and to examine the relationship of the latter variable to their acreage.

Most of the other counties on table 3.2 exhibit a reasonable relationship between real-property value and farm value. Not surprisingly, a large number of farm operators in the older Piedmont counties of Baldwin, Greene, and Wilkes seem to have owned considerable property outside their county of residence.[25] The number of such persons was lower in poorer regions characterized by small holdings: in the mountains (Towns and Habersham), in the upper Piedmont (Gwinnett), in the Wiregrass (Coffee, but not, for the reasons we have adduced, Emanuel), and in the Limesink (Worth). More problematic are the Central Cotton Belt counties of Baker, Clay, Houston, and Twiggs, which display strikingly different proportions of cases in which the value of real property exceeds farm value. The figure of 18.6 percent for Twiggs, a county in the older part of the belt, seems plausible; but in Houston, just to the west, the figure falls to 8.7 percent, lower than one would expect. In the southwest we would have expected relatively low percentages, since this was a region of recent settlement where many of the wealthiest proprietors were absentees who resided elsewhere. Yet the proportion of cases in Baker reporting excess real-property value is 20.3 percent, comparable with counties on older lands. More realistic for the newer lands is Clay's 6.8 percent. In that county five of the twelve type 1 operators with real property valued at $12,000 or more reported excess real-property value, in all cases by substantial amounts. The high percentage for Baker may have resulted from the loss of more than half its territory to newly created counties in the late 1850s. Some farmers who resided in the remnant of Baker might thus have "lost" some of their holdings to the new counties. The farm value for these holdings could have been reported on schedule IV in those counties rather than in Baker. The proportion of farm operators with excess real-property value also seems too high at 28 percent in Thomas, a county in the cotton producing upland region along the Florida border, though county formation in the 1850s may again account for some of the cases. The figure of 15.6 percent is also high in Echols, a relatively poor county on the palmetto flats near the Okefenokee swamp. While we have not attempted to reach more definitive conclusions about the real-property value–farm value relationship in these doubtful counties, our analysis of the problem in Harris and Emanuel will alert the researcher to the possibility that in any county the data may not in fact represent what it purports to represent.

Enough has been said to establish the possibility that real property was occasionally, or even systematically in some counties, over- or undervalued—probably the latter in most cases where it occurs. These faulty reportings could introduce serious distortions in the data that cannot yet be estimated. It is also clear that such distortions are most likely to occur and be significant precisely in those counties where multiple-county holdings were most significant and the real-property variable is of most interest: the cotton belt. In the case of farm value, we have seen that at least in Emanuel County it was highly defective as well. Finally, and more seriously, most enumerators failed to report farm value for tenants. This problem is discussed in connection with improved and unimproved acres in the next section.

Improved Acres and Unimproved Acres

Data on acreage can be important for testing some conflicting hypotheses about tenancy in the antebellum South. Most historians who have acknowledged the existence of tenancy have assumed that most tenants were "poor whites," eking out a living on poor lands on the fringes of plantations. More recently, Roger Ransom and Richard Sutch have suggested that "most" tenancy, at least in the cotton South, "involved the leasing of plantations or medium-scale farms that were operated by the leaseholder on the plantation system."[26] Acreage data, at least in counties where it was reported for tenants, can provide a measure for comparing the size of tenanted and owner-operated farms in various regions, although it cannot fully resolve all the questions raised by the "poor white" and "plantation" hypotheses. Farm value, cotton production, and personal property are also important variables that can be examined in this connection.

The census instructions defined improved acreage as that which is "cleared land used for grazing, grass, or tillage, or which is now fallow, connected with or belonging to the farm." It defined unimproved as "a wood lot, or other land at some distance, but owned in connection with the farm, the timber or range of which is used for farm purposes." It was not necessary that either improved or unimproved acreage constitute contiguous parcels of land so long as they were "owned or managed" by the person identified as the farm operator on schedule IV. So far the instructions seem to have excluded from the unimproved category wild lands, owned either by residents or nonresidents of the county, that were totally unconnected to the farming operation. The instructions went on to say, however, that the land "embraced under the two heads 'improved and unimproved' includes the whole number of acres owned by the proprietor" and only explicitly excluded "irreclaimable marshes of great extent" and "bodies of water of greater extent than 10 acres."

There is evidence that most enumerators resolved these ambiguities in favor of including as farm acreage all lands owned by farm operators in a county, with the possible exception of town properties. Map 3.1 and appendix E give the proportion of total farm acres from the 1860 aggregate census as a percentage of county area, taken from the 1960 census, for those counties whose boundaries remained substantially unaltered during that time. It can be seen that counties in the lower Piedmont and the upper Central Cotton Belt reported, with a few exceptions, a total farm acreage that nearly equalled county area. Farm acreage almost uniformly constituted a much smaller proportion of area in those parts of the state, the mountains and coastal

Map 3.1. Farm Acres as Percentage of Total Surface Area

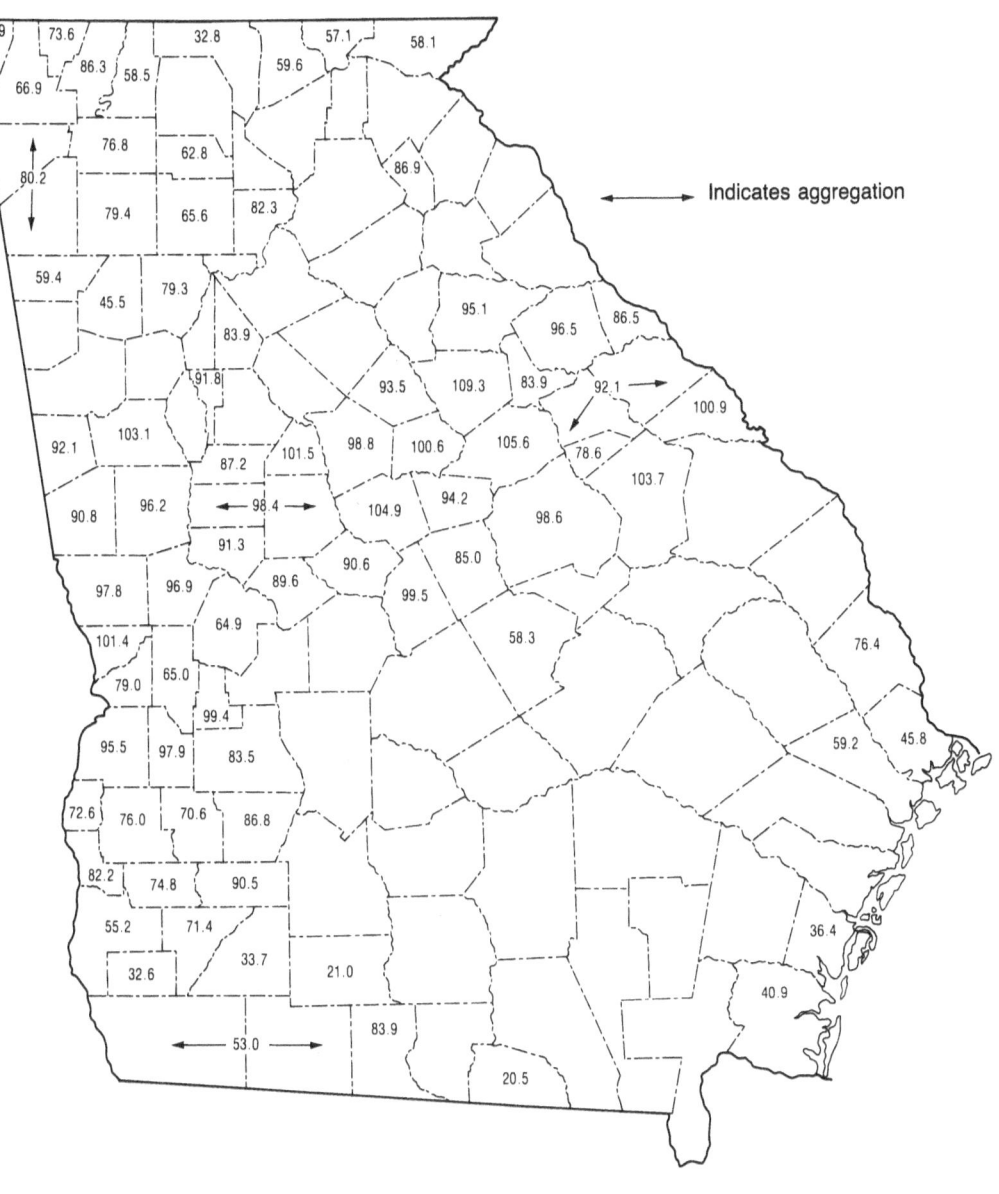

Notes: State mean is 70.7. For the data, see appendix E.

region, for example, where topography and soil conditions restricted farming and in the newer cotton lands of the southwest. Unowned or unclaimed lands or tracts held for speculative purposes might have accounted for much of the difference between farm acreage and area in these agriculturally marginal or less settled regions of the state.[27] Nevertheless, a good deal of uncertainty remains as to what sorts of land were considered by enumerators to be part of a farm. We cannot even exclude the possibility that some enumerators entered on schedule IV *all* farm lands belonging to residents of a county, even if some of the land was located outside that county. The census instructions did not explicitly prohibit such a practice, and they may even have encouraged it by specifying that noncontiguous parcels of land should be aggregated. But our previous discussion comparing farm value and real-property value does suggest that enumerators normally entered only the acreage located in the county.[28]

In Baldwin County data, we found an implausible relationship between improved acres and cotton production that raised additional questions about the reliability of the acreage data. In each of our sample counties we calculated the ratio of improved acres to bales of cotton for farms producing that crop—an admittedly crude indicator of cotton intensity.[29] Table 3.5 shows that Baldwin had the lowest ratio, hence the greatest cotton intensity, of any county in our sample. This is a surprising result, because most of Baldwin is located on what by 1860 were the relatively exhausted soils of the eastern Piedmont, with the southern part of the county consisting of the even less productive soils of the Sand Hills. A comparison of Baldwin's cotton-intensity ratio with those of neighboring counties, as shown on the bottom of table 3.5, is equally striking.[30] It may be that Baldwin farmers devoted an unusually large proportion of their acreage to cotton or followed unusually sound management practices. The other alternatives are that the enumerator either overcounted cotton bales or undercounted improved acres. The latter possibility is more likely since Baldwin, as table 3.5 shows, reported the lowest proportion of improved acreage of any major cotton county in our sample and also the lowest proportion among its neighbors.[31]

An underenumeration of improved acreage in Baldwin can perhaps be explained by the enumerator's failure to include in that category all "land which," as the census instructions put it, "has been reclaimed from a state of nature, and which continues to be reclaimed

Table 3.5

IMPROVED ACRES PER BALE OF COTTON AND AS PERCENTAGE
OF TOTAL FARM ACRES, SIXTEEN SAMPLE COUNTIES, 1860,
AND BALDWIN AND NEIGHBORING COUNTIES, AGGREGATE CENSUS, 1860

County	Improved Acres/ Bale[a]	Improved Acres as Percentage of Total Acres
Sixteen Sample Counties		
Baker	6.1	35.5
Baldwin	6.0	29.3
Clay	7.4	33.5
Coffee	14.0	4.7
Echols	16.8	12.8
Emanuel	26.6	7.6
Greene	13.4	43.5
Gwinnett	22.1	28.1
Habersham	48.6	14.6
Harris	10.3	52.5
Houston	6.1	49.3
Thomas	10.9	32.7
Towns	—	21.1
Twiggs	7.5	44.4
Wilkes	15.0	44.6
Worth	11.7	16.2
Baldwin and Neighboring Counties		
Baldwin[b]	6.4	27.5
Hancock[b]	8.3	33.9
Jones[b]	18.1	64.2
Putnam[b]	11.3	56.8
Washington[c]	11.7	34.3
Wilkinson[c]	8.7	37.9
Bibb[d]	9.8	40.6
Glascock[d]	15.2	34.1

[a]Cotton producers only in sixteen county sample; all farms in Baldwin and neighboring counties.
[b]Piedmont
[c]Central Cotton Belt
[d]Sand Hills

and used for the purposes of production." Improved acreage was to comprise grazing, grass, and fallow land as well as tilled land. Conceivably the Baldwin enumerator confined himself to recording only tilled acreage. One can also imagine some confusion among enumerators about whether to include *all* formerly cleared land in the improved category or to try to make distinctions between land actually lying fallow and old fields permanently withdrawn from production. Baldwin appears to be a fairly obvious case of underreporting improved acreage, but inconsistent practices in defining such acreage may have been more widespread, thus introducing another problematic element into comparative analysis.

The use of data on improved and unimproved acres, as well as farm value, in a study of tenancy in Georgia is limited by the fact that enumerators followed conventions that usually excluded one or more—and frequently all three (our standard convention)—of these schedule IV variables for tenants.[32] Unquestionably tenants with arable production, and the great majority of type 2a tenants did report arable, must have had improved acres, and the acreage they farmed must have had a cash value. They need not, however, have held any unimproved acres. Since it was not uncommon for enumerators in some counties to enter only improved acres for tenants, an examination of only these counties (in addition to counties reporting no acreage) might have led us to conclude that tenants typically leased only improved land, possibly with the right to let their animals graze or forage on the landlord's unimproved acreage. But two very different counties in our sample, Wilkes, in the old Piedmont cotton belt, and Worth, in the livestock-raising Limesink, both regularly reported unimproved acres for tenants.[33] The question, then, is, what happened to the missing acreage and farm value that the tenancy conventions failed to report?

The two possibilities are that these variables simply went unreported and therefore the acreage and farm value in many counties are under-enumerated, or that these variables were entered on schedule IV under the landlord's name. We have good reason to believe that the latter was the case. The widespread use of tenancy conventions suggests some degree of consistency in enumerators' methods, not necessarily in terms of the particular conventions employed but in the use of conventions as such, that cannot be attributed to whim or carelessness. The care some enumerators took in specifically identifying as tenants farm operators entered in any number of the different conventions again implies that these enumerators intended to show real distinctions between tenants and proprietors—and, as we will argue in chapter 5, possibly between different kinds of tenants. Unless enumerators failed to record somewhere all acreage in farms, together with its value in the district or county for which they were responsible, it is plausible that they would have assigned these amounts to the owners of the property. Under Georgia law the owner of land was responsible for taxes assessed upon it, and tax returns could be made only by the owner or his legal agent.[34] But the arable and livestock production of a rented farm belonged to the tenant.[35] Following common legal practice, then, an enumerator could have attributed a tenant's production to the tenant himself and assigned the acreage and farm value to the landlord. We are not suggesting that enumerators behaved like tax collectors or collected taxes at any point in their careers. We are suggesting, however, that the process of enumeration was occurring within an environment that had been conditioned by annual tax collections. On the other hand, some enumerators might have simply assigned the acreage and value to the farm itself, regardless of who worked it, and thus have ignored legal distinctions.

But what of those cases in which the enumerator assigned only one or two of the three variables to the tenant? An enumerator might have reasoned that since the tenant worked the land, the acreage should be assigned to him, but the land's value should be assigned to the landlord who paid the taxes. Yet such reasoning does not account for enumerators who recorded improved acres for tenants but not unimproved, or the enumerator in Emanuel County who entered only farm value. Once again ambiguity in the census instructions might help explain different procedures. It will be recalled that the name entered on schedule IV as the farm operator could be that of the "owner, agent, or tenant." The instructions for improved acreage referred to that "*owned or managed* by the person whose name is inserted in the [operator] column."[36] But in the case of unimproved acreage the instructions referred to such land "belonging to each *proprietor.*" The next sentence, however, defined total farm acreage as "the whole number of acres *owned by the proprietor.*" The instructions for farm value make no references at all to proprietors or anyone else. With such inconsistent language in the instructions, it is no wonder that enumerators were also inconsistent among themselves.

Both logic, based on the legal status of the tenant, and perhaps some confusion arising from the census instructions about how to assign acreage and farm value, point to the same conclusion: "missing" tenant values were aggregated with those of the proprietor. But data from Towns County provide more direct evidence. The Towns enumerator adopted the standard convention for all of the 172 type 2 tenants, who constituted 40.8 percent of the farm operators in the county. Livestock and crop output were assigned to the tenants themselves. Although the mean improved acreage for the 241 owner-operators (type 1) was only 56.2, 11 of these operators reported 200 or more improved acres and owned 29.4 percent of the county's improved acreage. But most of these 11 farmers reported strikingly small amounts of livestock and arable production. As table 3.6 shows, their share of every arable crop in the county produced in more than negligible amounts, and their share of all types of livestock except asses and mules, was markedly less than their share of improved acres, unimproved acres, or farm value. This result could perhaps be explained by the failure of the enumerator to enter tenanted acreage anywhere on schedule IV, thereby driving the acreage share of the 11 largest operators upward. This does not appear to have been the case, however. If we generously assume that the mean size of tenanted farms was twenty-five improved acres, then tenants would have held a total of 4,300 improved acres. If we further assume that tenants' acreage is missing from the returns and add their estimated acreage to the reported county total, we achieve 17,536 improved acres. Even with these assumptions the 11 largest proprietors would have owned 22.2 percent of the improved acreage in the county, a still greatly disproportionate share with respect to their arable production.[37] The alternatives are clear: either the large holders of Towns were experiencing severe *dis*economies of scale, or, much more likely, their reported acreage included a substantial amount of land farmed by their tenants.

In most counties, even where the proportion of tenants within the farming population was high, total tenanted acreage seems to have accounted for only a very small proportion of the total farm acreage. It is therefore not possible to offer additional tests, based on comparisons of total farm acreage reported in the census to the surface area of counties, that tenanted acreage and farm value, when not assigned to the tenant, were assigned to the proprietor. Map 3.1 shows

Table 3.6

COMPARISON WITH COUNTY AGGREGATES OF SELECTED VARIABLES FOR HOLDERS OF 200 OR MORE IMPROVED ACRES, TOWNS COUNTY, 1860

	Holders of 200 or More Improved Acres	County Aggregate	200 or More as Percentage of Aggregate
Improved Acres	3,890	13,236	29.4
Unimproved Acres	9,690	49,613	19.5
Farm Value ($)	72,400	260,042	27.8
Horses	64	654	9.8
Asses and Mules	35	148	23.6
All Cattle	287	2,701	10.6
Sheep	253	2,854	8.9
Swine	706	8,103	8.7
Wheat (bu.)	677	6,479	10.4
Rye (bu.)	730	4,576	15.9
Indian Corn (bu.)	10,300	98,571	10.4
Oats (bu.)	795	4,437	17.9
Tobacco (lbs.)	158	3,048	5.2
Peas/Beans (bu.)	14	866	1.6
Irish Potatoes (bu.)	87	3,449	2.5
Sweet Potatoes (bu.)	470	7,633	6.2
Value Orchard Prod. ($)	—	1,020	0.0

Note: This table includes data for all crops produced in more than trivial amounts. Towns reported no cotton production in 1860.

that total farm acreage was roughly equal to surface area in many counties. If we take Twiggs, where 34.8 percent of schedule IV operators were type 2 tenants in standard convention, and reported farm acreage was 99.5 percent of the surface area, we might anticipate further support for our assumption. If tenanted acreage had been missing from the returns and not assigned to proprietors, one might have expected total farm acres to have been significantly less than surface area. But if we assume that mean farm size for tenants was 80 improved and unimproved acres, or even a generous 100 acres, such acreage would have accounted for respectively only 5.6 or 7.0 percent of total farm acreage in the county. These figures, as can be seen from the map, are well within a reasonable margin of error between total farm acres and surface area.[38] These results for Twiggs, while certainly consistent with our assumption about the distribution of tenanted acreage, are far from conclusive. Nevertheless, both logic and the strong direct evidence from Towns County still lead us to conclude that enumerators almost certainly assigned missing tenant values to proprietors or their equivalents. The latter would have included managers or agents entered on schedule IV and classified as type 6 when the name of the proprietor did not appear as a separate record on schedule I.

A difficulty arises in connection with farms that may have been operated *exclusively* by tenants. In such cases there would have been no justification for the enumerator to make a separate entry on schedule IV specifying the owner's acreage and farm value, since there would have been no production to report under the owner's name.[39] It is possible that some of our type 3 entries, operators with at least improved acres and farm value on schedule IV but reporting no real property on schedule I, may have been instances of exclusively tenanted holdings for which the enumerator reported values in lieu of absent proprietors. But this suggestion is speculative, and if type 3 entries do not account for exclusively tenanted holdings, the possibility remains that acreage and farm value on such holdings went unreported.

The systematic nonreporting of variables for tenants in many counties not only restricts the kinds of analysis of tenancy that one would ideally like to pursue; it also poses serious difficulties for other uses of the census. If our assumption regarding aggregation of acreage and farm value under proprietors is correct, then the relationship between the production data for owner-operators and their reported acreage and farm value could be distorted in those counties where one or more of these variables was not recorded for tenants. The entry for an owner-operator on schedule IV would show only his own production but, if he had tenants, would overestimate to an unknowable extent the acreage, and the value of that acreage, on which his production took place. In many counties, as we have seen, tenanted acreage may not have had a significant impact on county aggregates. But much econometric analysis, using manuscript census data, has focused on such problems as economies of scale and the relative efficiency of slave labor on farms in different size classes. We have no way of knowing how tenanted acreage was distributed among such classes. If this acreage was disproportionately concentrated in one or more classes (however they may be constructed), its impact could have been considerable. Fogel and Engerman, for example, have attempted to measure the relative efficiency of southern agriculture in 1860 by using census data to relate the output of individual farms to their factor inputs of land, labor, and capital.[40] But since we have no way of knowing which proprietors held tenanted acreage, their conclusions may actually be enhanced by an upward bias in their acreage data, thus resulting in an overestimation of the productivity of labor and capital with respect to land.[41] The extent to which their conclusions are enhanced, however, depends upon the extent of the bias, which may or may not be large.

There are, then, severe problems in using farm acreage data from the census. Many enumerators did not record acreage for tenants; and if they assigned it to proprietors, as seems to have been the case, the relationship between the proprietor's acreage and production variables will be distorted to the extent that the magnitude of the acreage is large. Moreover, there is some reason to believe that not all enumerators defined either improved or unimproved acreage in the same way, and in no county can one determine with any degree of certainty how these definitions were applied. While the relationship of farm acreage to surface area in most counties where it could be tested seems reasonable, some anomalies do appear. In short, the acreage data, like other census data, must be used with caution, and their limitations must be recognized.

Personal Property

Personal property, recorded on schedule I, is the only census variable that provides a ready measure of the relative economic status of tenants among geographic regions. It is also the only census variable that permits consistent economic comparisons between tenants and other members of the agricultural population. "Personal Estate," according to the census instructions, was "to include the value of all the property, possessions, or wealth of each individual which is not embraced in the [value of real estate], consist of what it may; the value of bonds, mortgages, notes, slaves, live stock, plate, jewels, or furniture." It is clearly an indispensable variable for testing the leading alternative hypotheses about the status of southern tenants: whether they were mainly poor whites or leaseholders of large farms operated on the plantation system. The usefulness of personal property is enhanced by the failure of many enumerators to record acreage and farm value data for tenants.

The greater part of personal property, at least in the staple-producing regions, consisted of slaves. Table 3.7 compares the total value of personal estate in our sample counties with the estimated value of slave property based on the mean value of slaves in each county as reported by the state comptroller general and the total number of slaves as reported by the census. It can be seen that in all but five counties the value of slaves exceeded that of all other forms of personal estate. The figures, however, demonstrate the importance of slaves as a component of personal property in only the grossest kind

Table 3.7

COMPARISON OF AGGREGATE PERSONAL PROPERTY AND ESTIMATED VALUE OF SLAVE PROPERTY, SIXTEEN SAMPLE COUNTIES, 1860

County	Personal Property Value ($)	Slaves	Mean Slave Value ($)	Estimated Slave Value ($)	Slave Value as Percentage of Personal Property
Baker	2,149,390	3,492	731	2,552,652	118.8
Baldwin	8,388,435	4,929	697	3,435,513	41.0
Clay	2,638,932	2,253	740	1,667,220	63.2
Coffee	806,274	663	559	370,617	46.0
Echols	435,755	314	631	198,134	45.5
Emanuel	1,500,268	1,294	641	829,454	55.3
Greene	6,732,620	8,398	650	5,458,700	81.1
Gwinnett	3,227,362	2,551	693	1,767,843	54.8
Habersham	1,179,774	787	533	419,471	35.6
Harris	8,041,484	7,736	706	5,461,616	67.9
Houston	10,222,548	10,755	699	7,517,745	73.5
Thomas	5,522,562	6,244	657	4,102,308	74.3
Towns	234,690	108	638	68,904	29.4
Twiggs	4,960,053	5,318	665	3,536,470	71.3
Wilkes	8,989,512	7,953	672	5,344,416	59.4
Worth	728,967	632	721	455,672	62.5

Source: Mean slave value is taken from the *Annual Report of the Comptroller General of the State of Georgia*, 1860 (Milledgeville, 1860), 41–42. The other values are taken from the aggregate census for 1860.

of way and can in fact be highly misleading; for the *total* value of an individual's personal property was enumerated by the census in the county of residence whereas the number of his slaves was recorded in the county where "they are usually held to service." In other words, a resident of Baldwin who held slaves on a plantation in Baker would have had the value of those slaves entered as a component of his personal property in the former county but would have had the number of those slaves entered in the latter county. Such a hypothetical situation could easily have been duplicated in reality. Baker had a large number of absentee proprietors (who appear in our data as type 6), constituting 16.5 percent of total schedule IV entries and owning 39.6 percent of the county's improved acreage. It is therefore not surprising that the estimated value of slaves actually exceeded the value of personal property in Baker, since a large number of the slaves must have belonged to type 6 absentee proprietors. On the other hand, there was one owner-operator in Baldwin who reported $1,136,650 of personal estate and $388,040 of real estate, of which the value of his Baldwin farm accounted for only $6,800. Almost certainly he held most of his slaves as well as his real property in another county, probably in more than one.

There may be no necessary relationship between the value of an individual's personal property and the number of slaves he reported on the slave schedule of the enumerators' returns (schedule II). First, slaves held in other counties were not entered on schedule II for the county of the slave owner's residence, a circumstance that could lead to an undercounting of an individual's total slaves. Second, since slave hiring was a widespread practice in the antebellum South, slaves entered on schedule II may in many cases have been hired or rented by the slaveholder named on the schedule. This circumstance could lead both to an undercounting of the slaves owned by the persons hiring them out and a false attribution of ownership to persons who were hiring them. While a few enumerators explicitly identified owners and hirers on the slave schedule, most did not.[42] The census returns may offer a reliable count of the slave labor force that an individual had at his disposal on a farm in a particular county, but they cannot tell us how much that individual had invested in slave capital. Personal property in fact offers a better guide to a person's nonlanded capital resources and at least to his *potential* for slave ownership.

There is no reliable alternative data source against which the county census aggregates of personal property can be tested. The comptroller general of Georgia did report the value of personal property by county in 1860. However, unlike for the census, the value of slave property was supposed to be assessed for taxation in the county where the slaves were held, not in the county of the slaveholder's residence. Thus the census returns and the comptroller general's report are not comparable in this respect. It is possible that some census enumerators disregarded their instructions and reported only the value of personal property held in the county of residence. Table 3.7 shows that the estimated value of slave property in Greene County, for example, accounted for 81.1 percent of personal property, the highest percentage in any of our sample counties except Baker. Greene is located in the older lands of the eastern lower Piedmont, where one would expect many of the wealthiest planters in particular to have invested in land and held slaves elsewhere. To the extent that this was so, the slave percentage of personal property ought to have been driven downward, since the value of slaves in the denominator would have included slaves held outside the county, while the value of slaves in the numerator would have been restricted to those held in the county. The slave percentage of personal property in Greene's neighbor, Wilkes, was only 59.4 percent, a more reasonable figure for the region. Similarly, the older Piedmont county of Baldwin reported less than half of the value of its personal property in slaves actually held in the county.[43]

There is also evidence that some enumerators may have been less than accurate in their reporting of personal property, particularly for those of relatively little wealth. Table 3.8 provides a frequency distribution for all persons in our sixteen-county sample reporting personal property valued at less than $500. In most of the counties, personal property was reported in each of the categories below $100, but there are some anomalies, at least among those records that appear in the data base for this study. Only three persons in Baldwin had no personal property, a strikingly smaller proportion than in the older cotton counties of Greene and Wilkes. Moreover, all three were type 3 records that also reported no real property. These cases could easily have been errors for which both values were inadvertantly left out. In Echols County no one reported personal property under $100, with the exception of three who reported none; once again, all were type 3 records. In both counties it seems that the enumerator assumed

Table 3.8

FREQUENCY DISTRIBUTION FOR PERSONAL PROPERTY OF TYPES 1a, 2a, 3a, 4, 5, AND 7a REPORTING $0–$499, SIXTEEN SAMPLE COUNTIES, 1860

County	$0		$1–$49		$50–$99		$100–$499	
	No.	%	No.	%	No.	%	No.	%
Baker	14	23.0	2	3.3	—	0.0	45	73.8
Baldwin	3[a]	3.6	4	4.8	10	11.9	67	79.8
Clay	13	9.9	2	1.5	7	5.3	109	83.2
Coffee	2	1.4	12	8.2	22	15.1	110	75.3
Echols	3[a]	11.1	—	0.0	—	0.0	24	88.9
Emanuel	13	5.7	7	3.1	16	7.1	190	84.1
Greene	35	38.9	1	1.1	3	3.3	51	56.7
Gwinnett	34	4.8	40	5.6	70	9.8	567	79.7
Habersham	18	3.1	32	5.6	78	13.6	446	77.7
Harris	23	9.8	8	3.4	11	4.7	193	82.1
Houston	27	19.3	4	2.8	6	4.3	103	73.6
Thomas	93	43.9	2	0.9	3	1.4	114	53.8
Towns	36	11.4	42	13.2	46	14.5	193	60.9
Twiggs	11	6.8	3	1.8	23	14.2	125	77.2
Wilkes	32	42.1	—	0.0	1	1.3	43	56.6
Worth	4	3.5	2	1.8	6	5.3	102	89.5

[a]Includes type 3a only.

everyone must have had *some* personal property; but, whereas the Baldwin enumerator apparently attempted to provide a more or less accurate measure of its value if it fell below $100, the Echols enumerator simply assigned an arbitrary minimum of $100. On the whole, however, regional patterns are apparent in table 3.8, suggesting, if not accuracy, at least consistency in reporting. The distributions in the two oldest Piedmont counties, Greene and Wilkes, are remarkably similar.[44] Habersham, a mountain county on the lower ridge toward the Piedmont, is very close in its distribution to Gwinnett in the upper Piedmont. Towns, a more isolated county on the far ridge, not unexpectedly shows a higher proportion of cases in the two lowest categories. The two neighboring Central Cotton Belt counties of Houston and Twiggs show some differences in their distributions, but not enough to raise serious questions about them. The Wiregrass counties of Emanuel and Coffee and the Limesink county of Worth also have distributions that are not strikingly different from each other. The only other anomaly is Thomas on the recently settled Florida border whose distribution is much closer to Wilkes and Greene than to nearby Baker in the Central Cotton Belt or nearby Worth in the Limesink. But of the cases in Thomas reporting no personal property, almost 80 percent were type 5, "farmers without farms" and without real property. Thomas County, as we will argue in chapter 5, may have had an unusually large number of sharecroppers, reported as type 5—as opposed to renters, reported as type 2—among its tenants. Sharecroppers might be expected to have owned very little, if any, personal property.

On the whole, it appears that a fair degree of confidence may be placed in the consistency of reporting personal property. But, as with other variables, the data in every county must be considered on their merits. The Greene enumerator may have reported personal property held only in that county. In addition, some enumerators may not have attempted accurate estimates of personal property, particularly for those of relatively little wealth; such a practice, while not having much impact on the overall distribution of *most* personal property in any county, can seriously distort the pattern of wealth-holding for precisely those persons whom this study addresses: tenants.

Bales of Cotton

The only production variable that we have included in our data base is the number of 400-pound bales of ginned cotton produced by each farm operator. We can therefore not measure differences in crop mix between various categories of the agricultural population or between different regions, although such measurements are important to a full understanding of the place of tenancy in the agricultural economy. Gavin Wright, for example, has argued that to avoid risk small-farm operators in the cotton south strove first for self-sufficiency and devoted the balance of their resources to the production of a marketable surplus.[45] It will be important to know whether tenants followed the apparent antebellum practices of small operators generally or prefigured postbellum patterns of increased concentration on cotton at the expense of subsistence. To the extent that tenants were not self-sufficient, they must have depended to some extent on credit facilities to purchase supplies. Did the landlord provide the credit (or perhaps food surpluses from his own land), or did the tenant rely on country merchants who, as Ransom and Sutch argue for the postbellum period, may have locked him into cotton production? These and other questions, especially those relating to the nature of antebellum leasing and sharecropping arrangements, will ultimately have to be answered to determine whether tenancy after the Civil War represented a radically new departure or the elaboration of earlier trends.[46]

Our use of the cotton variable can provide partial answers to other questions about antebellum tenancy. The proportion of the total crop produced by tenants in each county or district will show the degree to which the cotton economy depended on tenancy in various regions. The ratio of improved acres, where available for tenants, to bales of cotton will offer a crude measure of differences in the degree of "cotton intensity" among the different categories of farm operators.[47] This is hardly a satisfactory surrogate for a measure of crop mix, but it will be at least a first approximation of the relative extent to which tenants were engaged in the production of the major staple and involved in a market economy. Finally, since we have chosen counties within and without the major cotton-producing regions, we will be able to examine in a provisional way the relationship between dependency on cotton and tenancy rates.

Unless enumerators regularly standardized the production of each farm's cotton output into 400-pound bales, bias of unknown magnitude may be introduced into computations using cotton as a variable. That they did standardize bales is problematic. It is well known that cotton bales varied widely in weight in the antebellum South.[48] There was no consistency at all in the weight of the bales farmers marketed. Few enumerators may have taken the trouble to convert the pounds of cotton produced by a farmer into the standard census bale but may have recorded the number of bales, regardless of weight, into which a farmer had pressed his cotton. If this was the case, we can achieve only the roughest approximation of the relative importance of cotton on any farm or within any category of farm operator. Refined econometric analysis may be impaired to the extent that such analysis is dependent on reliable magnitudes.

4

The Delineation of Agricultural Regions in Georgia

SOCIAL AND ECONOMIC HISTORIANS have increasingly come to place great importance on soil and climate as prime variables in explaining spatial variations in social structure and patterns of production. Moreover, most historical analysis of landholding in the southern United States, as well as much of the traditional wisdom and folklore, have relied heavily upon generalizations about regional differences within the various states. For example, historiographical traditions associated with antebellum Georgia expect to find substantial differences between the exhausted old lands of east Georgia and the fresh new lands of the western cotton belt with respect to crop productivity and mix, proprietor residence and absentee patterns, and ratios of blacks to white population. Similarly, before the Civil War, one would expect to find the upper Piedmont and mountain counties, as well as the Wiregrass region, characterized by smallholding white owner-occupiers, dirt poor in the mountain and Wiregrass regions, and having relatively low levels of black population. These expectations have always been at least implicitly based on considerations of soil and climate and on the known patterns and chronology of land settlement. It is therefore at least heuristically useful to examine rates of tenancy within antebellum Georgia in terms of their distribution within a system of agricultural regions defined roughly by soil, climate, and chronology of settlement. A more sophisticated mathematical analysis may also wish to treat soil, climate, and date of settlement in a more serious way as independent variables with precise values. We have not undertaken so elaborate a procedure, but we shall attempt to draw sharp distinctions between analysis at the two levels and to note briefly the obstacles that bar the more sophisticated treatment.

All discussions of agricultural regions within the southern states during the antebellum and postbellum decades have been largely based on data and categorization presented in the 1880 United States Census Office report on cotton production.[1] That report divided Georgia into seven large agricultural regions, each region being defined principally by its dominant soil and climatic characteristics, and proceeded to offer a wealth of detail relating specific soil and climate data to patterns of production. The regions were very broadly defined, several being divided into two or more subregions or divisions for separate treatment. The most notable of the subregion distinctions was that made between the Limesink division and the Pine Barrens division of the Long Leaf Pine and Wiregrass region. The 1880 cotton report, along with rank-order statistical analysis, has formed the basis for the most recent formulation of southern agricultural regions by Roger Ransom and Richard Sutch. They have modified the 1880 cotton report scheme for Georgia by creating one additional region in the northwest (comprising Dade County); by separating the Sand Hills into a separate region instead of leaving it as a subregion of the Central Cotton Belt; by substantially redrawing the boundaries of the Limesink division and treating the two Wiregrass divisions as separate major regions; and by creating a third new region (comprising Echols and Charlton counties) delineating an area where sea island, rather than the usual short staple, cotton was grown. The Ransom and Sutch scheme, displayed on map 4.1, is a considerable refinement and improvement of the scheme found in the 1880 cotton report if one considers only the broad outline of major regions. But we shall find that the 1880 cotton report was considerably more sensitive in the details of its discussion and provided grounds for delineating a number of additional subregions that might substantially advance spatial analysis. We shall return to the 1880 report and to the Ransom and Sutch scheme after examining briefly the major physical characteristics of each of the agricultural regions within the state. The descriptions that follow are based principally on the 1880 report.

Both the 1880 cotton report and Ransom and Sutch distinguish two major regions among the mountainous counties of northern Georgia: the Ridge and Valley region in the northwest, and the Blue Ridge region in the northeast. Northwest Georgia is a region characterized

Map 4.1. Sixteen Sample Counties Placed within the Economic Regions of Georgia

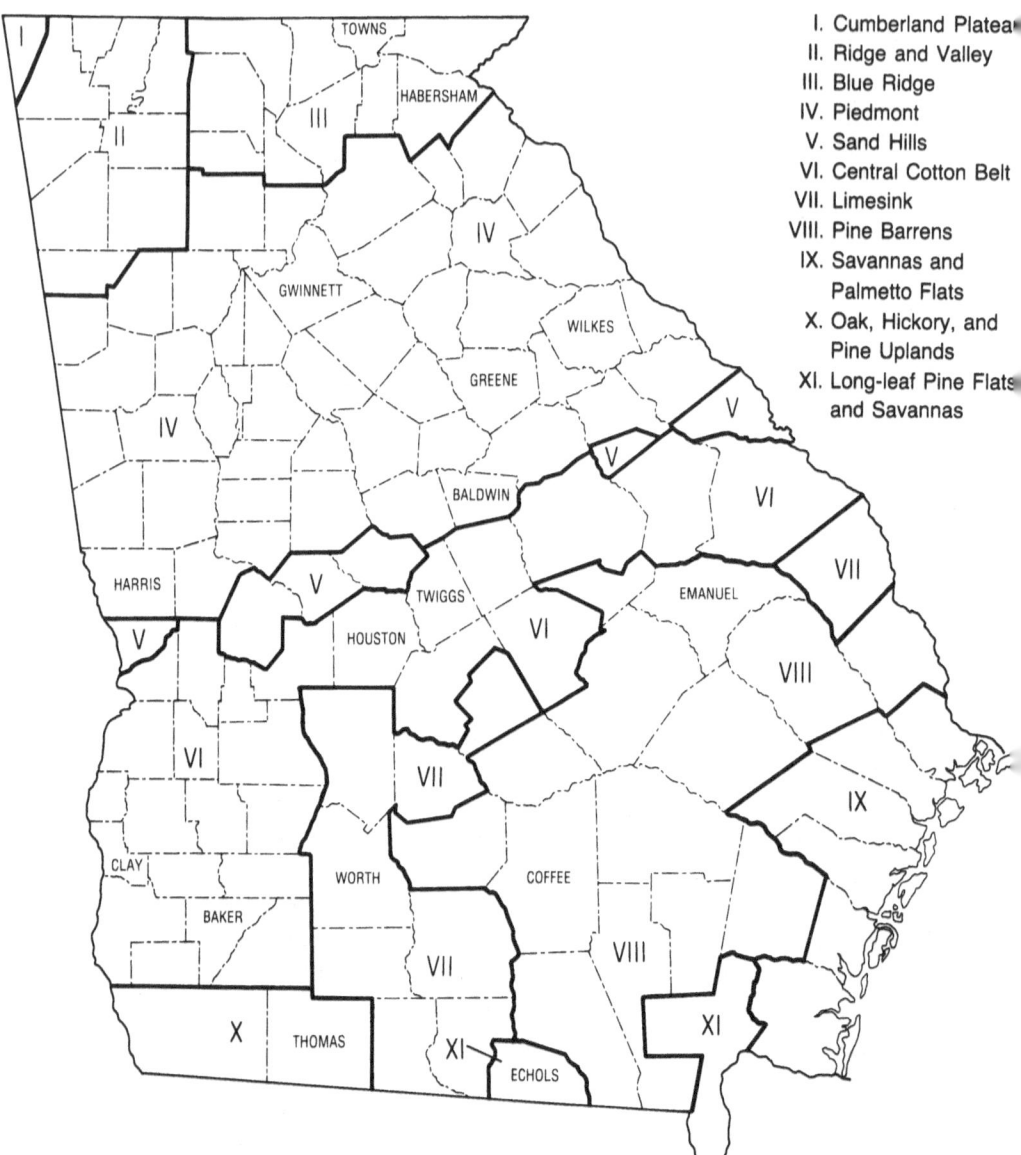

I. Cumberland Plateau
II. Ridge and Valley
III. Blue Ridge
IV. Piedmont
V. Sand Hills
VI. Central Cotton Belt
VII. Limesink
VIII. Pine Barrens
IX. Savannas and Palmetto Flats
X. Oak, Hickory, and Pine Uplands
XI. Long-leaf Pine Flats and Savannas

Source: The economic regions are the Roger L. Ransom and Richard Sutch modifications of the 1880 Cotton Report (*Economic Regions of the South*, Working Paper No. 3, Southern Economic History Project).

by high abrupt ridges running generally in a south-southwesterly direction with broad agricultural valleys lying between them. The soils in these valleys are extraordinarily diverse and intermixed; and while the floors of the valleys vary only from 500 to 1,000 feet above sea level, the range of the variation is agriculturally significant. Some of these valleys often have favorable soils for arable cultivation but are located at such an altitude that early frost and short planting seasons severely limited their suitability for many row crops before the introduction of massive fertilization in the late nineteenth century was able to substantially shorten the time it took for the crops to mature. Nevertheless, while some of the valleys were certainly too high for successful planting in the antebellum period, a proper perspective is more clearly gained if one remembers that the city of Atlanta, in Fulton County in the upper Piedmont, is at an elevation of 1,035 feet. Arable production within the region was also stimulated when the Western and Atlantic railroad linking Atlanta and Chattanooga was constructed through the heart of the area in the late 1840s. The Ridge and Valley region, then, should be considered as one of great agricultural potential, largely unrealized in the antebellum period, but already in process of rapidly accelerating structural change.[2]

The Blue Ridge counties of northeastern Georgia contain the highest ridges and mountains of the state; and while interspersed with wide agricultural valleys, these valleys are at an average elevation of 1,600 to 1,800 feet above sea level, an elevation that is too severe for the profitable cultivation of many row crops, even under conditions of modern fertilization. Even in 1879, after intensive fertilization had begun, Fannin and Towns were the only counties in the state to report no cotton production whatever, while other counties in the region produced the staple only in small patches for home consumption. Moreover, until the late nineteenth century, the region was unpenetrated by efficient transport systems and was virtually cut off from the market economy, such as it was, of the upper Piedmont; indeed, even the river systems on the far side of the ridge flowed northward into the remote areas of North Carolina rather than south into Georgia. The region was characterized by subsistence farmers surviving on small plots along the water courses and engaged in nonstaple production and livestock husbandry. They enjoyed a reputation for illiteracy and independence, and in many ways they were culturally distinct from the rest of the state. They were backwoodsmen whose families had migrated into the region along the Appalachian chain from the mountainous regions of Virginia and the Carolinas, and they were more similar to the peoples of the Wiregrass than they were to those of the central cotton economy.[3]

As soon as one descends from the Blue Ridge, even into the hilly mining counties of the extreme northern Piedmont, the land becomes more favorable to cultivation. Whereas one-third of the Blue Ridge region is too broken for cultivation, only 7 percent of the eleven hill counties north of the Chattahoochee River is too broken; when one passes below the Chattahoochee, the figure is reduced to about 1.5 percent. The upper and lower Piedmont is an east-west belt about 70 miles wide, descending gradually from just over 1,000 feet to about 600 feet above sea level on its southern boundary at the edge of the Sand Hills. The soils of the region are principally gray sands and red clays, occurring in belts of every width and patches of every size. The wide varieties of parent material and the relative lack of transporting agencies prevented the blending of soils found in the more southerly portions of the state below the fall line, a narrow belt of hills running in a half circular swing through several southern states. But while the soils are highly interspersed, there is some tendency for them to run principally in belts northeast to southwest, as do the geological formations from which they derive. Prime cotton lands thus tend to cut through portions of counties in highly discontinuous belts, making simple classifications of counties by soil more than usually difficult. The lands of the eastern Piedmont were among the earlier settled in the state, and many had been under cotton cultivation since the late eighteenth and early nineteenth centuries. The lands of the lower middle Piedmont, by contrast, were settled and brought under cotton only in the second two decades of the nineteenth century, while the lower western Piedmont counties were brought under intensive cotton production only in the 1820s and 1830s. The upper Piedmont counties, because of their higher elevations, were not so intensively given over to cotton before the introduction of heavy fertilization in the decades following the Civil War. There are, thus, sharp distinctions to be found within the Piedmont region, between the older and highly exhausted lands of the east, the intermediate lands of the middle lower Piedmont, the fresher lands of the lower western Piedmont, and the fresher and less staple-oriented counties of the middle and western upper Piedmont. This last area, unlike the others, was never fully integrated into the staple- and slave-based economy of the lower and eastern Piedmont before the Civil War. One thus expects the upper Piedmont counties to be characterized by white middle to small holders, with large cotton plantations being principally concentrated in other areas of the region. Traditional wisdom also leads one to expect to find many planters on the older lands continuing to reside on the old homestead but transferring much of their capital, in the form of slaves and new land acquisitions, to the fresher and now more productive lands of west and southwest Georgia. For these older counties, the real and personal wealth variables on schedule I of the census should in many cases substantially exceed values found on the agricultural and slave schedules.

The 1880 cotton report divides the Central Cotton Belt into three distinct subregions: the Sand Hills, abutting on the Piedmont on their northern boundary and comprising a narrow continuous belt of highly broken and eroded acidic sand running at various widths—from a few miles to over 20 in Taylor and Marion counties—from the Savannah to the Chattahoochee rivers; the Red Hills, an even narrower belt abutting on the Sand Hills in the east but dropping away southward and tending to disperse as they reach the more western counties, always being found in discontinuous belts and patches, especially in the west; and the Yellow Loam belt, which comprises the remainder of the region. The soil of the Sand Hills was among the worst of the state. It occurs on the fall line, which separates the higher elevations of the southern Piedmont regions from the lower coastal plain. River systems running into the coastal plain are generally navigable only to the fall line, where the sudden rise in elevation tends to produce cataracts and falls. Many major southern cities are therefore located on the periphery or within the boundaries of the Sand Hills; Columbus, Macon, and Augusta provide examples in Georgia. The soils of the Sand Hills are mildly productive when fresh, but they are acidic and highly eroded, their natural bases having been leeched out and with little or no alluvial deposit allowed by the swiftly moving fall-line currents. They are thus quickly exhausted when subjected to intensive cultivation, especially when under cotton. The Sand Hills are geographically in the heart of cotton and plantation country but are only marginal participants in the cotton economy. Farmers on these lands may be expected to be smallholders engaged in low levels of staple production, probably under the stimulus of local markets and immediately adjacent production patterns, but more basically engaged in subsistence agriculture.

The Red Hills, which lie interspersed to the south of the Sand Hills, had a high rolling and well-timbered surface, principally in hardwoods indicating a high base content and high fertility. The soil has a sandy surface but a rich clay subsoil. East of the Flint River, which runs through Macon County, these red lands are generally large in contiguous area, highly productive, durable, and easily tilled. They were very good for cotton, but farmers tended to prefer them for small grains. West of the Flint, the red lands are more sandy and, while productive, are quickly exhausted. In the extreme northwest of the region, they lie on terrain that is too broken for cultivation, farmers preferring the yellow loam of the valleys.

The best cotton lands of the region, and the center of the late antebellum plantation economy, were to be found on the Yellow Loam, which occupied the majority of the surface area, extending from the Savannah River on the east to the Chattahoochee on the west and as far south as Early County. To the east of the Flint River the Yellow Loam country was rolling, and its high base content and fertility was indicated by a cover of oak, hickory, and, especially, long-leaf pine. The soil was gray sand on the surface, with a clay subsoil. Throughout the region, but more notable in some counties than in others, there were to be found rich deposits of marl and limestone, sometimes in beds at the surface. It was the presence or absence of these bases that were more important than anything else in determining whether the unusually acid and eroded Ultisols of the southeastern United States were to be fertile for the production of row crops. Particularly rich beds of marl and limestone were found in Washington, Houston, and Dougherty counties. While the Yellow Loam seems, judging from the 1880 report, to be uniformly good to the east of the Flint River, the quality of the soil appears highly mixed and regionalized as one moves west and south. Soil quality seems to improve markedly as one moves southward through the valleys and onto more level land, the soils becoming progressively less sandy and more loamy with clays nearer the surface and having a higher incidence of and softer quality of limestone subsurface.

As they were in the Piedmont, the patterns of settlement within the Central Cotton Belt are highly varied. The eastern counties were already under cotton in the late eighteenth and early nineteenth centuries. The middle counties were settled and brought under cotton during the second two decades of the nineteenth century. The extreme southwest began to be settled in the 1820s and the upper western counties only in the 1830s; but throughout the western and southwestern portions of the region the most important clearances and expansion into cotton occurred during the 1840s and 1850s, as the older lands to the east became progressively exhausted. Much of the capital for clearance and development came from planters whose older lands to the east were largely abandoned or, as it was termed, "lying out." They acquired new fresh lands in the fertile southwest and sent many of their slaves to work them. But the planters themselves frequently stayed home on the old plantations, partly for reasons of sentiment, but also in part because sections of the southwest had a nasty reputation for being unhealthy and disease ridden. One should therefore expect insalubrious counties such as Baker to have a high incidence of absentee proprietors (our type 6), while counties, such as Early, enjoying a finer reputation, should have a higher rate of residence. Cotton intensity, slave concentrations, and the incidence of large holdings should to some degree follow patterns of soil and clearance.

The Long Leaf Pine and Wiregrass region was divided by the 1880 cotton report into two distinct divisions: the Pine Barrens and the Limesink. Ransom and Sutch retained and further refined that distinction. The Pine Barrens covers the greater part of southeastern Georgia, comprising an area approximately 100 miles wide. Its soil is almost uniformly highly acidic sand, lacking a clay subsoil and seriously deficient in the bases required for the successful cultivation of most row crops. As in the Sand Hills, the soil of the Pine Barrens may give reasonable yields for a short period, but they are quickly exhausted. The rivers of the region provided little bottom lands where a richer alluvial might form. By the first decade of the twentieth century a new wave of farmers had settled into the region, and by introducing massive fertilization they were able to achieve high levels of cotton production; yet around 1880 the cotton report insisted that the region "cannot properly be called a cotton-growing section of the state." Some 6 percent of the area was judged to be irreclaimable swamp (principally, one assumes, in the vicinity of the Okefenokee), while of the remainder only 15 percent had been cleared for cultivation. If one remembers that the practice of the region was to abandon land as it became exhausted in favor of new clearance, it can be seen that arable production did not dominate the agrarian regime. The 1880 report noted that a "large percentage" of the land cleared was then lying out and only about 5 percent was at that time under actual cultivation. The farmers of the region were in fact backwoodsmen engaged in subsistence arable production that was subordinate to a pastoral regime of animal husbandry, supplemented by hunting and fishing. Turpentine production and lumbering were significant industries within the region, especially in the vicinity of navigable waterways. Population densities in the antebellum era were among the lowest in the state.

The Limesink division superficially resembled much of the Pine Barrens. Its surface soil was for the most part the same acidic sand found further to the south and east. But the Limesink enjoyed two considerable advantages: the region was underlaid by soft limestone subsoils and often had shell rock in the surface soil, which provided the bases needed for cultivation; and the rivers of the division cut generous bottoms and secondary hummocks. It was in these bottoms and hummocks that cotton production of the Limesink was principally concentrated. These bottoms consisted of a dark alluvial soil, about 10 to 20 inches deep, well supplied with bases, and a clay subsoil. They were durable and produced high cotton yields, but, as with all bottom lands, excessive moisture made cotton subject to rust, and many farmers preferred corn, for which these lands were ideally suited. A glance at the 1859 agricultural returns of the division will quickly reveal that cotton was a feature of the agrarian regime, but not a dominant one, most farmers being engaged in lumbering and turpentine production and in mixed husbandry, with considerable emphasis on livestock. Notable differences in production patterns between individual farmers probably reflect differences between holdings on the sandy uplands and the loamy bottoms. One would expect the division to be characterized by middle to small holders with relatively low levels of capital formation. Much of the area was yet to be cleared. As late as 1880 only one-fifth of the land in Worth County had ever been cleared for cultivation, the remainder still being covered with forest.

The two or three counties in the extreme southwestern corner of the state—comprising Decatur, Thomas, and Brooks counties—form a segment of a separate soil region that reaches south and westward into Florida and is called the Southern Oak, Hickory, and Pine Uplands. The area is higher and more rolling than the Wiregrass which, according to the 1880 report, abuts on it as Limesink on the north and east.[4] Its surface soil is a relatively thin sand underlaid with clay and lime and is moderately productive. A peculiar feature of the region is the red clay loam found in small localities where the timber growth is oak and hickory. These soils are favored for arable production. The region, like the Limesink, is also favored with rich hummocks, but, again, rust is a problem for cotton production. The region is basically favorable to mixed husbandry, with some emphasis on grazing, but cotton was moving increasingly into the region in the later antebellum years. The region had a more healthful reputation than its low-lying neighbors, so long as one stayed away from the rivers, and one would expect to find relatively high rates of residence among a farming population that was still characterized principally by middle to small holders.

The extreme southeastern corner of the state is dominated by the vast Okefenokee swamp, which is principally in Ware, Pierce, and Charlton counties. The swamp itself is of no agricultural significance except for hunting and forage. On its periphery is a subregion described by the 1880 report as the Pine and Palmetto Flats, similar to the Savannas running north along the coast counties but at a higher elevation and more swampy. The soils are sand and bog and extend through most of Echols and Clinch counties on the west, northward through most of Ware and half of Pierce, and eastward into a small portion of Camden. Ransom and Sutch separate Echols and Charlton counties into an additional region, not because of distinctive soils but because these counties, like those of adjacent counties in northern Florida, produced principally sea island cotton rather than the usual short-staple strain. It is probable that the entire subregion should receive further special treatment. The area was only of marginal economic importance, however. The population density was thin indeed, the holdings small and ill-defined, with much squatting. In Echols County cotton was confined almost entirely to narrow oak and hickory belts running along Suwanoochee Creek and the Alapaha River. These bottom lands had a rich lime subsoil, a rarity in this sector of the state, and were reported in 1880 as rich, durable, and relatively well drained. The remaining soils of the county were principally acidic sand, thickly covered with saw palmetto. Clearance could be expected to be a major problem throughout the subregion.

The final region delineated by the 1880 cotton report lies almost entirely within the boundaries of the six coastal counties. Five subregions of agricultural importance are noted. One of the largest is the Savannas, lying in a long continuous belt between the Savannah and St. Mary's rivers about 10 to 15 miles wide and 10 to 15 feet above tidewater. This belt constitutes the first terrace in the rise from sea level, abutting on the Wiregrass to the west, where the second terrace begins. The Savannas occupy the greater part of Chatham, Bryan, Glynn, and Camden counties and large portions of Liberty and McIntosh, with a sizable portion of Liberty and a section of Bryan extending into the Pine Barrens. The Savannas are broad, flat, open plains, having no growth other than sparse and tall long-leaf pine and a thick undergrowth of saw palmetto. Both the surface and subsoils are sandy and highly acidic, and they were not much subject to cultivation before intensive fertilization was introduced later in the century. The rural economy was dominated by livestock and characterized by light capital formation. Population densities were thin throughout the region.

A second belt of Live Oak Lands extends throughout the region in an irregular and interrupted fashion. The soils of this belt are a yellow or mulatto sand and were recognizable to contemporaries by the large live oaks draped with spanish moss. The upland and intermediate lands of this belt were not favored for cultivation, but the bottoms were very rich, and short-staple cotton was grown.

The Coast Tide Swamp Land forms a third narrow discontinuous belt bordering on inlets and streams up to the limits of the tidewater. It was on these tidelands that the great rice plantations of the state were to be found. Georgia's few rice planters were immensely wealthy. Rice production was highly labor intensive; and since the intense heat and malarial conditions of these lands made residence by whites uncomfortable and life-threatening, the slave population ratios in four of the coastal counties were the highest in the state.

The 1880 report notes a number of small areas of Marsh Land at the mouths of some rivers but makes no agricultural comment on them. The fifth subregion consists of the Sea Islands themselves, which form a continuous interlocked belt 10 or 12 miles wide along the entire length of the Georgia coast. The islands have a sandy and acidic soil, but were well adapted to the production of sea island cotton, corn, and sweet potatoes, as well as fruit and olives. The planters on the islands were frequently the same persons as those engaged in rice production. The cultivation of sea island cotton, as that of rice, was highly labor intensive and wholly reliant on slavery. When slavery disappeared, so did large-scale staple production in the region.

It should already be evident by now that the construction of regional agricultural boundaries in Georgia is replete with difficulties. Not the least of these difficulties lies in an inconsistency between the actual boundaries of soil systems, and thus of agrarian regimes, and of the bulk of statistical data presently available on landholding, production, and demography. The boundaries of soil systems and agrarian regimes cut erratically across county boundaries, making some counties an almost hopeless mixture of soil types and production patterns. The boundary problems experienced by Ransom and Sutch, in restructuring the perimeters of the Georgia Limesink and the south central portions of the Central Cotton Belt, are due to the unusually complex patterns of soil associations in the southwest portion of the state and the fact that many counties there and elsewhere extend deeply into more than one major soil region. But data derived from the published aggregate census are not reported below the county level, and those using such data are perforce restricted to agricultural regions defined by county boundaries.

It is possible to overcome such restrictions to a degree, though only with an extraordinary infusion of additional labor, by constructing statistical series from the manuscript census schedules in those counties where the returns are defined by militia district or land district. County sizes can be great and can incorporate maximal variations in soil, whereas the relatively small sizes of militia and land districts yield a far more satisfactory level of homogeneity and permit sharp intra-county differences to emerge. The advantages of militia-district-level analyses, along with their own peculiar liabilities, will become clear when we examine three counties in this manner in chapter 6. But it remains unfortunately true, even if one were to decide that militia-district-level data were suitable, that not all counties were reported by militia or land district; and any procedure that was dependent on such districts would have to either include all counties where reporting by district occurs or restrict itself to a sample of such counties. The number and distribution of counties available would determine the decision. Both options would be dependent on a more detailed specification of soils than is available in the 1880 cotton report, which is more often than not too vague in describing their precise locations. Fortunately, more precise literature is becoming available in the form of 1-inch-scale and detailed highway maps produced in Georgia by the state Department of Transportation and in the slowly emerging series of county soil-survey booklets compiled by the Soil Conservation Service of the United States Department of Agriculture. The general highway maps offer details on water courses and precisely delineate the boundaries of militia districts. In this imperfect world, however, the researcher may expect to find that militia district boundaries have themselves been altered and new districts created in all those counties affected by new county creations subsequent to the period of investigation. For this reason we have found highway maps to be valuable in most regions of Georgia but of little use in much of the mountains and Wiregrass and some portions of the upper Piedmont. Land districts may be reconstructed from nineteenth-century county maps, where available, but also at a more generalized scale on nineteenth-century state maps. Their boundaries are further described in the legislation that established the land lotteries early in the century. Land districts are not found in those portions of the state organized under the headright system before 1803. In those areas, however, relatively little boundary change occurred.

The soil survey booklets, along with large-scale Soil Association and Land Use Potential maps also produced by the Soil Conservation Service as part of their soil survey booklets, provide a wealth of detail that is both helpful and full of pitfalls for the historian. The information is precise, giving details of soil analysis for all varieties encountered; the booklets also contain a map series that provides a bird's-eye view of soil configurations at both the associational and subassociational levels. Statistics are provided that permit percentage estimates for each soil association within each county. The difficulty in using such data lies in estimating fertility potentials for each soil association under various conditions of land use. The booklets and maps offer generalized ratings for row crops, but these ratings, as well as the discussion surrounding them, are related to conditions pertinent to the mid-twentieth century, under conditions of modern fertilization and soil management, and after a century of further erosion and nutrient depletion or alteration.

Soils of the Ultisol order, which characterize most soil types found throughout the southern United States, pose special problems for analysis.

> These are the most weathered of all the midlatitude soils and are found on surfaces that are dominantly Pleistocene or older. Glacial ice did not rest on these soils during the Pleistocene period. Consequently, they have been subject to weathering processes for a much greater length of time than have the soils of humid regimes on their northern boundaries. Extensive chemical weathering of their solum [surface soil] has led to a removal of bases; yet contrary to expected increased nutrient availability with depth, the reverse is true. . . . "the maintenance of bases in the surface horizon is at the expense of the supply in the deeper horizons." Once the native plant cover is removed, as for planting crops, the meager store of nutrients is rapidly lost and potential crop yields decrease dramatically. Only through conscientious fertilization programs can permanent agriculture be practiced on such soils.[5]

Apart from an increasing use of guano, only traditional methods of marling and grazing were in widespread use in the South before the Civil War. It is precisely for this reason that natural sources of base, such as lime and marl, were so important in the prime cotton lands during the antebellum period. Only in the later decades of the nineteenth century, especially toward the turn of the century, does modern intensive fertilization begin in the American South. If southern soils had experienced glaciation, as had those of many northern states and much of Europe, it might be possible to focus on the glacial drift, which as a subsoil would then act as a relatively uneroded parent material, especially where the drift was not so sandy as to permit high levels of leeching. Such an option is not available in the South.[6]

The options for the southern states would seem to be either to accept highly generalized fertility indicators derived from the 1880 cotton report, as occasionally modified by comments and estimates given by the Soil Conservation Service, or to construct a precisely valued index of soil fertility based on certain currently measurable contents. For example, it may be possible to construct a fertility index for premodern soils of the Ultisol order by using modern data on "percentage base saturation" (PBS), which measures the relative amount of exchangeable bases contained in a soil type, and the pH scale, which measures the relative concentration of hydrogen cations and thereby determines relative acidity. The pH scale is peculiarly relevant to the highly acidic soils of the South. If one were to couple such an index with area percentage estimates of each soil association or subassociation within each county, it may be possible to incorporate soil as a precise variable in mathematical analyses of southern production patterns. The application of such an index, along with area percentages for each soil subassociational grouping, may also make it possible to undertake mathematically precise analysis incorporating soil.

Such mathematical analyses, while pertinent to many questions touching on tenancy, go well beyond the interpretive objectives of this volume. Our present objectives are more modest: to establish the substantial presence of tenancy in all or many of its varieties in antebellum Georgia, to explore its relationship to the cotton economy, and to examine whether the distribution of tenancy displayed meaningful relations with soil and other production variables and with patterns of settlement. What is needed for these purposes is a regional structure that is sufficient to support generalization and to facilitate discussion. We intend that the discussion proceed more by inspection than by statistical analysis in any strict sense and that a flexible understanding of variations in the countryside be applied. Any numerical values we offer are of necessity provisional, and so must be our interpretation of such values. Our sixteen-county sample (map 4.1) is not presented as a "sample" in any rigorous sense, and as such it requires no statistical defense. For our purposes, then, we are content to employ a roughly defined system of regional boundaries based on counties, and our discussion will base itself on the agricultural regions defined by Ransom and Sutch. But in employing such a scheme we shall give careful attention to subregions between which we expect to find substantial differences in rates of tenancy; and in fact many hypotheses cannot be tested without the employment of subregions. We shall therefore give some attention to differences between the older and newer lands of the eastern, middle, and western lower Piedmont, with separate attention given to the middle and western upper Piedmont. Differences in tenancy rates should flow gradually as one moves southward from the Blue Ridge to the fall line and eastward from the Alabama border to the Savannah River. Similarly, within the Central Cotton Belt, considerable differences should occur as one moves westward onto the newer lands, with some complexity occurring in the southwest. The sixteen-county sample was designed merely to test the water for such flows.

Some of the counties included in the sample (Gwinnett, Worth, and

Thomas) were selected not only for their regional locations but also because they formed part of the 1880 sample of southern farms drawn by Ransom and Sutch. Worth also had the considerable advantage in 1860 of reporting full values for tenants on schedule IV and of labeling tenants as such, while Gwinnett was reported by militia district. Houston was selected because of its production characteristics, but its reporting proved so erratic and its values so low that Twiggs, another Ransom and Sutch county and its neighbor on the east, was added to the sample. Baker was included to test the controversial Limesink area of the southwest and to provide a transition from Thomas into the principal area of the southwestern Central Cotton Belt. Some counties were frankly selected because of their small size: Clay in the southwest, Echols in the Palmetto Flats, and Towns in the Blue Ridge. When Towns produced such high tenancy rates, we added Habersham in order to test the plausibility of such values and to examine the possibility of a downward flow in tenancy rates as one descended along the near slopes of the Blue Ridge into the hill country. Greene was added when Wilkes produced such low values on the old lands of the eastern Piedmont. Both Baldwin and Twiggs had the advantage of testing the districts in which they principally lay, but they were both reported by militia district, with some districts of each county lying in the Sand Hills. Coffee was selected when Worth produced unexpectedly high values when compared with Emanuel and because it reported improved acres for most tenants. It also provided an interesting transition from the Limesink into the newer lands of the Pine Barrens. Emanuel was included as our representative for the Pine Barrens on older lands for two reasons: Tattnall, selected by Ransom and Sutch, appeared to us at the time to have no tenants, and subsequent investigation has led us to place it in class 4;[7] whereas Emanuel quite evidently had tenants, taking the highly unusual procedure of reporting for them only their farm value. This piqued our curiosity. As for Harris, it was the first county we subjected to the long method. It was our pilot—where we learned there was something to be learned, and where we initially developed our method. It was districted, and appeared to be well reported; and we felt that the newer lands of the lower western Piedmont would be a good place to begin looking for tenancy in the Cotton South. It was.

5

The Forms of Tenancy in Antebellum Georgia

WE ARE ABLE TO DEMONSTRATE in this study that high rates of tenancy among the white agricultural population of Georgia did not have to await the dislocations caused by the Civil War, the emancipation of the slaves, and the proliferation of a new marketing and credit structure after 1865. A substantial number of white farmers did not own their own land in 1860. A more complex set of issues, however, concerns the forms of antebellum tenancy and whether they anticipated postbellum patterns and, most important, whether sharecropping and renting, were practiced to a significant extent before the Civil War. Harold D. Woodman and Steven Hahn, while admitting that some amount of tenancy had existed in the antebellum South, have insisted that "the cropper relationship" was almost entirely a postwar phenomenon and that a new form of what was in effect wage labor, absorbing both whites and blacks into a new proletariat, replaced the slave regime.[1]

In this chapter we will not undertake a definitive discussion of these issues. Our aims are more restricted. We propose to see what light an examination of the census returns themselves might shed on forms of tenancy in antebellum Georgia. We will show that enumerator procedures are suggestive of a variety of tenancy arrangements. More cautiously, we will argue that the procedures of some enumerators are at least consistent with the presence of sharecropping. Finally, since even a simple count of tenants in any county depends on an interpretation of varying enumerator procedures, we will suggest a method for estimating a range of tenancy rates that incorporates at each of four levels different assumptions about which type of census entry may be identified as a tenant.

The forms of postbellum southern tenancy were exceedingly complex, and as a result there has been a good deal of confusion both among contemporaries and later historians about the meaning and implications of different tenancy arrangements. Various terms such as sharecropping (or cropping), cash renting, share renting, renting on thirds and fourths, and standing rent abounded. Most historians (including the present writers) conventionally use the term "tenant" to include both renters and croppers; but in strictly legal terms only renters of land were true tenants, whereas croppers, even though they typically farmed separate and distinct parcels of land, were laborers for whom their share of the crop constituted the wage payment. The renter maintained control of the crops; the cropper did not. The renter worked free of the landlord's daily supervision and control, although the rental contract might have specified in some detail which crops were to be grown, rotation patterns, and various tasks the lessee was to undertake on the farm, such as fence-mending or clearance. The cropper, by contrast, was subject to the continuous supervision of his employer. Sharecropping must not in turn be confounded with share renting (or share tenancy), in which the tenant agreed to pay the landlord a fractional share of the crop as rent.

Other forms of renting included cash renting, in which the payment was a fixed amount of money, and standing rent, in which the payment was a stipulated quantity of the output rather than a fractional share. A type of share renting that was apparently common mainly among whites in the hill country and mountains, where relatively less cotton was produced, was the thirds and fourths system. Under this arrangement the tenant paid out one-third of the grain and one-fourth of the cotton. By the 1870s the legislatures or courts of every southern state had attempted to formalize the distinction between landlord-renter and employer-cropper, but even so, the courts had difficulty in deciding particular cases when the words of the lease or contract were ambiguous or when, as was often the case, there was no written agreement between the parties at all. Moreover, the looseness in terminology did not help matters; for, as Charles S. Mangum, Jr., has pointed out, "In ordinary parlance [a share renter] may sometimes be referred to as a 'sharecropper.'"[2]

For the historian attempting to unravel the tenancy knot, the census office compounded the problem when it began in 1880 to distinguish tenants from proprietors. In doing so it grouped tenants into two misleading categories: "rents for fixed money rental" and "rents for share of products." Strictly speaking, sharecropping does not belong under either of these rubrics, since the cropper was not a renter but an employee. Many students of postbellum tenancy have ignored the distinction between sharecropping and share renting and the ambiguous status of the census categories.[3] It is likely that some enumerators, who were aware of the legal distinctions, entered only cash renters and share renters on the agricultural schedule, excluding sharecroppers altogether. Most probably lumped croppers and share renters into the same category, but in this case the proportion of one type to the other simply cannot be determined. Map 7.2, which shows the proportion of tenants who were "cash renters" in Georgia's counties in 1880, does display some coherent regional patterns, but the markedly varying proportions of cash renters in some counties of the same region also suggest that some enumerators differed among themselves in classifying tenants.[4] Robert Preston Brooks, a special census agent studying plantation conditions in 1911, believed that most white tenants in the upper Piedmont and mountain regions of Georgia were renters on thirds and fourths while most blacks in those regions were croppers but that both groups had been placed in the share renter category. "Some confusion in the statistics" was the result. Brooks even wondered whether the census should have considered the holdings of croppers as separate farms at all.[5] It is apparent, then, that the census tenancy categories for 1880 and beyond may be highly misleading. Before that year the census did not classify tenants at all, but there are a number of indications in the earlier returns that many enumerators were aware of different forms of tenancy and that the complexities of agricultural tenure and labor were just as great in 1860 as in 1880.

The difficulties that planters faced after the Civil War in reorganizing agricultural production on the basis of free labor and the time it took them to experiment with various combinations of gang labor, renting, and sharecropping have led some historians to assume that they were largely unfamiliar with the practice of family tenancies, particularly sharecropping.[6] It is likely, however, that the period of uncertainty and experimentation that followed the war resulted not so much from a lack of familiarity with family tenancies as such but, on the one hand, from the desire of planters to retain control over their labor force and, on the other, from the desire of the freedmen to maintain as much independence as possible from planter control. A perusal of the *Southern Cultivator,* a leading agricultural journal published in Georgia, for the years after 1865 reveals these planter concerns as well as the skepticism of many about the capacity of blacks to function on an independent basis. To think that blacks could work on their own, declared one correspondent, was "utopian." On the other hand, some correspondents favored one or another kind of tenancy and were more or less optimistic about blacks' abilities.[7] Although the *Cultivator* made no specific references to prewar tenancy practices, there is no indication in that journal that planters were unfamiliar with family tenancies in general or with sharecropping in particular. One does get the sense that planters were casting about among many possible options for coping with an enormous labor problem. Tenancy as such was not unprecedented. What was unprecedented was the quantitative jump in tenancy rates after 1865 and the transformation of most of the four million slaves into wage laborers, renters, and croppers. This fact and the desire of planters to maintain their control over labor—and not necessarily the emergence of entirely new forms of tenancy—may have been responsible for the efforts of southern judges and legislators in the 1870s to clarify in court decisions and statute law a murky legal situation.[8]

The bulk of direct evidence that historians have so far accumulated on forms of antebellum tenancy does suggest that renting for cash, a fixed quantity of output, or on shares predominated, but the data are spotty and hardly conclusive. In his analysis of fifty-nine antebellum tenancy contracts from Hayward County in western North Carolina, Joseph D. Reid, Jr., has attempted to make a case for continuity into the postbellum period. Unfortunately he has confused share renting with sharecropping, using the latter term when in fact describing the former. All the contracts he has cited were clearly either for a standing rent (which Reid has taken to be equivalent to cash renting), share rental, rental in exchange for clearance, or a combination of these types. Sharecropping in the strict sense does not appear in his data.[9] Steven Hahn's examination of tenancy contracts for antebellum Carroll and Jackson counties in Georgia has found no evidence of sharecropping, but numerous instances of cash renting and share renting (including the thirds and fourths system). "Perhaps more frequently," Hahn has suggested, "planters rented out tracts of land to have them cleared and improved." Only a few landlords furnished supplies to their tenants, at least as far as surviving contracts reveal.[10]

The only unambiguous evidence so far adduced for sharecropping in the antebellum South comes from North Carolina court records. The courts there recognized the fundamental legal distinction between croppers and renters at least as early as 1837. In that year the state supreme court ruled that

> if a man engages another person to come and labour on his farm, as overseer or cropper, and stipulates with him that he will have a share of the crop for his labour and attention, the property in the entire crop is in the employer until the share of the overseer, or cropper is separated from the general mass.... Before the separation the labourer's right rests upon an executory contract with the employer. Before separation it could not be levied on to satisfy the labourer's debts.

The cropper was thus considered to be the employee of the landlord, with no property right in the crop. Significantly, the courts recognized an "implication" of cropping whenever, in the absence of a written lease to the contrary, the arrangement was on shares and the landlord furnished provisions.[11] Whether the courts in antebellum Georgia and other southern states also maintained these legal distinctions must await further research. Yet there is little reason to believe that sharecropping should have been confined to North Carolina; for, with the exception of large-scale tobacco production, the kinds of agricultural regimes in that state were largely present in Georgia.

Certainly the postbellum rulings of the Georgia courts were strikingly similar to the North Carolina precedents. In 1872, for example, the supreme court of Georgia declared that

> there is an obvious distinction between a cropper and a tenant [i.e., renter]. One has possession of the premises, exclusive of the landlord, the other has not.... The possession of the land is with the owner as against the cropper. The case made in the record is not the case of a tenant. The owner of the land furnished the land and supplies. The share of the cropper was to remain on the land and to be subject to the advancement of the owner for supplies. *The case of the cropper is rather a mode of paying wages than a tenancy.* The title to the crop subject to the wages is in the owner of the land.[12]

At least by 1872, then, a cropper in Georgia, like his antebellum counterpart in North Carolina, was considered an employee, and a presumption of cropping derived from the furnishing of supplies by the landlord. Moreover, the early postbellum Georgia rulings rested on statutes that did not yet explicitly distinguish between cropper and renter. The law code did not formally make the distinction until 1889.[13] Therefore, the mere absence of statute law defining sharecropping in the antebellum period cannot be taken as evidence that the institution did not exist in Georgia at that time, although such an absence may suggest that it was of lesser importance.[14]

It might also be argued that if sharecropping had been practiced on some scale in antebellum Georgia and other southern states, leading agricultural journals would have made some mention of it. The *Southern Cultivator*, for one, was silent on the matter. But such silence, while suggestive, is not decisive. The *Cultivator* actually made no mention of tenancy at all, and not in any of its possible forms. Lewis Gray has made a case against the presence of substantial tenancy by pointing to the paucity of references to it in the standard antebellum sources. The evidence from the census returns, however, shows that tenancy was widespread, so in this case at least the literary evidence can be misleading. Regardless of other problems in evaluating the reliability of the census enumeration, the fact that several enumerators explicitly identified large numbers of tenants—whatever forms such tenancy might have taken—should leave little doubt, despite the literary evidence, of their substantial presence.[15]

The case against antebellum sharecropping relies, then, on the virtual absence of the usual sorts of evidence for it, in the form of surviving contracts, discussion in the agricultural journals, and so on. The lack of evidence, with the exception of the North Carolina court cases, is quite compelling; but, as we have seen, historians have also been misled by the sources about the significance of tenancy generally. The census enumerators' returns themselves provide some evidence for considerable variety in tenure forms. The case for sharecropping is more problematical, however. Certain peculiarities in enumerator procedure and some bits of more direct evidence in the returns are consistent with its presence, but the census data alone allow us only to engage in the sort of plausible conjecture that further research may or may not sustain.

Our argument for complexity in tenure arrangements is essentially a circumstantial one and depends upon the conviction that minimal patterns of intelligibility can be detected amidst the variety of enumerator practices. The inconsistent practices among two, three, or some small number of enumerators in the mode of entering and identifying members of the agricultural population certainly appeared unintelligible to us during the early stages of our research. But the comparison of enumerator procedures first in the sixteen counties examined by the long method and then in the larger number of counties examined by the short method convinced us that a limited number of more or less consistent patterns could be detected. Moreover, individual patterns were not regionally specific. Radically different patterns appeared in otherwise similar counties in the same region. While we recognize that enumerators may not have been particularly intelligent or clever in their choice of procedures and that many enumerators were careless and internally inconsistent, we do believe that the principal patterns themselves were not without their own kind of logic. The problem lies in interpreting what these patterns meant.

It is apparent, first of all, that enumerators used inconsistent criteria for establishing who was a farm operator to be entered as such on schedule IV. It has already been shown that some enumerators placed on that schedule not only genuine producers who met the census criterion of production valued at least at $100 but also first-year farmers with no arable production and other individuals, mainly town dwellers, with only minor livestock or no production of any kind (types b and c).[16] Furthermore, in most counties "farmers without farms," some reporting and some not reporting real property, were entered only on schedule I (types 4 and 5). In some counties both types 4 and 5 appear in substantial numbers, in other counties one type is either absent or scarcely present, and in a third group of counties both appear but only in very small numbers. The question is whether the different proportions of types 4 and 5 reflected real differences among counties, or whether some enumerators excluded from schedule IV *certain kinds* of farmers that other enumerators entered on that schedule. A similar question can be raised with respect to the reporting of farm laborers and other laborers in often very different proportions among counties on schedule I only. In the case of both farm laborers and farmers without farms, the evidence points toward inconsistent procedures. Finally, a multiplicity of enumerators' conventions was found on schedule IV with respect to the reporting of acreage and farm value for tenants. As appendix A shows, most enumerators were consistent, or nearly so, in employing a single convention, although these conventions differed among themselves. A large minority of enumerators, however, employed two or more conventions. Moreover, many enumerators explicitly identified tenants on either schedule I or schedule IV, in a few cases using two or more designations that suggest different forms of tenure. It is on the basis of these variations in procedure that complexity in tenure arrangements can be argued and a more tentative case for the existence of sharecropping proposed. A cumulative argument can be developed, no part of which may be convincing in isolation.

A few enumerators seem to have excluded all tenants from schedule IV. This practice was apparent only in the five class 5 counties of those analyzed by the short method.[17] Of these five, however, Effingham had an unusually high number of farmers without farms on schedule I only, reporting no real property (type 5). Clayton and Whitfield reported similarly large numbers of household heads who were "farm laborers" and appeared only on schedule I. Such entries, to judge from patterns found in other counties in their regions, were probably tenants. Only Monroe and Telfair produced few apparent candidates for tenants on schedule I.[18] It will be recalled that while the census instructions for schedule IV specified that the farm operator could be a "tenant," the form itself mentioned only the "owner, agent, or manager." Probably a few enumerators hastily overlooked the instructions and were guided instead by the heading of the farm operator column, thus failing to record tenants on schedule IV. Counties in class 5 present few opportunities for detecting multiple forms of tenancy.

The possibility of sharecropping is suggested in the first instance by the presence of farmers without farms (types 4 and 5) in counties that also report type 2 tenants in the normal fashion on both schedules. Historians have variously assumed that types 4 and 5 consisted of some combination of retired farmers, farmers passing through a county and caught in the census net, ordinary farm laborers (in the case of type 5), and residents of the county with their farm property in another county (in the case of type 4). All these possibilities doubtless have some validity, although few farmers without farms seem to have been of retirement age. Some number of type 5 could have been ordinary farm laborers, especially in those counties that reported no household heads with that occupation (see table 5.1). Seddie Cogswell has offered what is up to now the most plausible explanation for farmers without farms. He argues that most of these persons were first-year farmers who had recently settled on the land and had not yet made a crop which could have been harvested before the end of

98 Farm Tenancy and the Census in Antebellum Georgia

Table 5.1

TYPES B AND C AS PERCENTAGE OF TOTAL SCHEDULE IV ENTRIES AND RATIO OF TYPE 4, TYPE 5, FARM LABORER HEADS OF HOUSEHOLD, AND OTHER LABORER HEADS OF HOUSEHOLD TO TOTAL SCHEDULE IV ENTRIES, SIXTEEN SAMPLE COUNTIES, 1860

County	Types b and c[a] (%)	Type 4 Ratio (× 100)	Type 5 Ratio (× 100)	Farm Laborer Ratio[b] (× 100)	Other Laborer Ratio[c] (× 100)
Baker	—	6.1	15.1	7.5	1.4
Baldwin	0.3	8.0	2.2	1.6	15.1
Clay	—	9.2	8.5	3.9	0.4
Coffee	2.6	1.2	2.0	—	—
Echols	—	—	—	—	43.0
Emanuel	—	2.6	4.1	2.2	—
Greene	0.5	6.1	10.4	0.7	3.9
Gwinnett	—	9.0	12.8	8.7	3.7
Habersham	10.8	0.3	5.5	—	0.9
Harris	5.7	2.7	8.4	—	1.8
Houston	12.8	1.8	9.1	1.1	2.6
Thomas	—	22.9	44.0	—	4.5
Towns	4.8	0.2	1.4	—	—
Twiggs	7.2	—	4.9	—	3.2
Wilkes	—	6.0	5.0	18.3	—
Worth	—	2.4	2.1	9.4	4.2

[a]For a breakdown between types b and c, see table 2.1.
[b]Includes "farm laborer" and "farm hand."
[c]Includes "laborer," "common laborer," "day laborer," "hireling," and "workman."

the census year on June 1.[19] Unless they had held substantial numbers of livestock, they would not have met the census criterion, a minimum of $100 in output, for inclusion on schedule IV. At least no enumerator adhering to his instructions would have entered them on schedule IV. But some obviously did enter them on that schedule, reporting only livestock or no production. Many of these first-year farmers, almost certainly in every county where they were entered on schedule IV, appear in our data as types b and c.[20]

It is likely that most types 4 and 5 in counties producing no types b and c (or only a negligible number of them) were first-year farmers. Every county must have had some number of such farmers; and if an enumerator, faithful to his instructions, excluded them from schedule IV, where they would then not appear as types b and c, they must have appeared as types 4 and 5 on schedule I. Indeed, as table 5.1 shows, every county in our sample with negligible numbers of types b and c—with the exception of Echols—contained more than a negligible number of types 4 and 5.[21] There are, however, a number of counties in our sample, most obviously Habersham, Harris, Houston, and Twiggs, with *both* significant numbers of types b and c *and* significant numbers of type 5. The hypothesis that type 5 records in these counties were first-year farmers thus appears implausible.

We propose an alternative hypothesis that could explain such type 5 records. Some enumerators, at least those in the four counties just cited, might have entered *some kinds* of tenants on schedule I and IV (type 2) and *other kinds* of tenants on schedule I only (type 5). That such type 5 cases were not first-year farmers is supported by the presence on schedule IV of types b and c, which must themselves have included the first-year farmers. In addition, the same four counties reported relatively few type 4 farmers—none at all in Twiggs (see table 5.2). If the enumerator had entered first-year farmers on schedule I only, we would have expected a significant number of type 4 as well as type 5 farmers to have appeared on the returns.[22] A distinction between type 2 tenants and type 5 "tenants" could have corresponded to a distinction between renters and sharecroppers. In the minds of some enumerators, renters, who were legally in possession of the crops they produced, could have merited entry on schedule IV as the operators of their holdings and with their production reported while croppers, who were merely employees of the landlord, might not have. In some counties type 5 could have included both first-year tenant farmers of all varieties and sharecroppers who were not first-year farmers but whom the enumerator otherwise excluded from schedule IV. This may be the case in Thomas County, for example, which reported no types b and c but extraordinarily large numbers of types 4 and 5.

Our hypothesis that type 5 records may include sharecroppers must be tested in other regions, especially outside the South, where sharecropping is presumed to have been insignificant throughout the nineteenth century. If such regions produce counties that repeat the patterns exemplified by Habersham, Harris, Houston, and Twiggs, then our hypothesis must be reconsidered—or the question of sharecrop-

Table 5.2

TYPE 5 AS PERCENTAGE OF TOTAL TYPES 4 AND 5, SIXTEEN SAMPLE COUNTIES, 1860

County	Type 5
Baker	71.1
Baldwin	21.9
Clay	48.0
Coffee	63.6
Echols	—
Emanuel	61.1
Greene	63.2
Gwinnett	58.8
Habersham	94.1
Harris	75.8
Houston	83.1
Thomas	65.8
Towns	85.7
Twiggs	100.0
Wilkes	45.2
Worth	46.2

ping in regions where it had not been suspected may have to be reopened. A study headed by Richard A. Easterlin, using the Bateman-Foust sample of northern farm households, has found high proportions of farmers without farms. The Easterlin group agrees with Cogswell that some number of such persons may have been first-year farmers. Their data also show an increasing ratio of farmers without farms as one moves from older to newer lands. This finding reinforces the plausibility of the first-year-farmer hypothesis, which accords well with the facts in many of our counties. Since the Easterlin study does not provide a breakdown by county for the data or any information on the number of types b and c, it is not possible to determine whether any individual counties display patterns similar to those in our four counties.[23]

Whatever may have been the status of type 5 farmers in the four counties, there is clear evidence that enumerators could exclude some, but not all, tenants from the agricultural schedule. In our short-method analysis we found four counties, identified in column

13 of appendix A, that explicitly noted tenants on schedule I who were not traceable to schedule IV. In all four counties, other tenants, presumably with different characteristics, were entered on both schedules I and IV. Even more suggestive, the enumerator in Walton County designated twenty "croppers" in the occupation column of schedule I. None of them, as far as we could determine, appeared on schedule IV. In Paulding County, on the other hand, we found, out of ninety entries tested for the short method, one cropper on both schedules and one on schedule I only. The latter could have been a first-year farmer.

We have proposed that in the counties of Habersham, Harris, Houston, and Twiggs, all of which seem to have reported first-year farmers on schedule IV, the type 5 farmer without a farm may have been a sharecropper. There is a problem with this hypothesis, however. Only Houston reported farm laborers on schedule I only. The other three reported no farm laborers (see table 5.1). This may mean that the type 5 farmers in these three counties were ordinary farm laborers who were given an inappropriate occupational designation by the enumerator. But it is also possible that some of the individuals designated in these counties as "laborer," "day laborer," or "hireling" were in fact farm laborers. That such designations could be used for farm laborers is supported by the returns for Echols County, where the ratio to total schedule IV entries of "laborers" who were heads of household and entered only on schedule I is an astounding 43 percent. Echols was an overwhelmingly agricultural county and could not have sustained such a large nonagricultural labor force. The practice of the Echols enumerator, though suggestive, does not prove anything about other counties, especially in the face of widespread inconsistency generally among enumerators. Yet the term *farmer* does at least imply something more than a mere farm laborer. Moreover, the Houston example still stands. There the farm laborer ratio is not very different from that in the older cotton-belt counties of Baldwin and Greene.[24]

The large number of presumed farm laborers in Echols County implies that some enumerators confined certain kinds of tenants to schedule I. If we grant that the laborers there were engaged in agriculture, then it is unlikely that most of them were ordinary farm laborers. There were simply too many. In some counties examined by the short method we also found unusually large numbers of farm laborers, but such counties contrasted sharply with others in their region and produced relatively few tenants on schedule IV. By neither the long method nor the short method did we locate any *region* in which a proportion of farm laborers anything like that in Echols was characteristic across more than one or two counties. A more plausible explanation is that most of the laborers in Echols were tenants of some kind, possibly sharecroppers, whom the enumerator may have wished to distinguish from renters. If they were sharecroppers, their designation as laborers would have corresponded to their legal status. Moreover, the enumerator entered "rents land" in the real property column on schedule I for all type 2 tenants in the county, thus making a distinction between sharecroppers and renters even more pointed.

A different sort of county, as far as the farm laborer problem is concerned, is Clayton, in the upper Piedmont. The enumerator, as far as can be judged by the short method, entered virtually no tenants on schedule IV but did report, apparently on schedule I only, a large number of "farm laborers" who were heads of household. Indeed, out of a total of 552 households in Clayton, 115, or 20.8 percent, were headed by a farm laborer. In addition, there were twenty-one "day laborers" in the county. If we assume that the farm laborers in Clayton were tenants and add them to total schedule IV entries, their percentage of such entries is 28.2. This rate compares reasonably well with the long-method type 2 tenancy estimate of 20.1 percent in Gwinnett County, also located in the upper Piedmont. If Gwinnett's farmers without farms (type 5) are included in the tenancy estimate for that county, it becomes 27.8 percent, practically the same as the Clayton estimate. A number of other counties in the upper Piedmont are also close to the Clayton estimate (map 6.6). To conclude that the farm laborers in Clayton were not tenants would mean that the county occupied an anomalous position in its region. It would, however, be rash to conclude that any or all of these farm laborers were sharecroppers, because the short method revealed no other tenants recorded in a different fashion. The term *farm laborer* in Clayton appears to have been used to encompass tenants of all sorts, even though it makes little sense to use it for renters. Only when, as in Echols, the enumerator distinguished large numbers of farm laborers from other tenants can a plausible conjecture of sharecropping be made. It nevertheless remains possible that the laborers in Echols represented a form of tenancy other than sharecropping, perhaps share renters as opposed to cash renters; but then it is puzzling why renters on shares should be designated as laborers instead of as farmers.

Much work needs to be done on how farm labor was used in the antebellum South and to what extent it could imply a form of tenancy. Undoubtedly some labor was casual and hired by the day or task, especially during the peak harvest and cotton-picking season. After the Civil War mixed systems existed whereby some farm laborers worked under yearly contracts for a proprietor, being employed most of the time on his land while having use of separate parcels for their own subsistence. Such laborers could be inaptly designated day laborers.[25] In the antebellum period, farm labor could also be hired by the year, as the following comment by a northern Virginia writer in 1838 implies.

> There is to the common laborer in this country every temptation to idleness.... How can it be expected that he will remain stationary with his employer at $100 or $150 a year, when for twenty, he can buy an old horse and second-hand cart, or steal them, and after begging his way to Illinois and Wisconsin, can squat on a quarter section of land.[26]

Whether such "common laborers" were typically paid in cash or with a share of the crop—and whether they farmed separate parcels of land—bears heavily on the question of whether sharecropping of the classic postbellum type existed before the war.

The presence of enumerator conventions in reporting acreage and farm value may also relate to tenancy arrangements. The most fundamental distinction in law was between sharecropping on the one hand and renting (in its multifarious forms of cash renting, standing rent, and share renting) on the other. In the four counties where type 5 farmers could have been sharecroppers, all but a negligible number of certain tenants (type 2) were entered in standard convention, that is, their acreage and farm value were not recorded on schedule IV.[27] In legal terms such a distinction would have made sense, since the sharecropper owned neither the land nor the crops produced on it, while the renter owned the crops but not the land. In the former case (type 5) both crops and land could have been assigned to the owner on schedule IV, but in the latter case (type 2) the tenant could have received credit for the crops but not for the land or the value of the farm.

One difficulty with the hypothesis that type 5 farmers in some counties were sharecroppers has already been pointed out: they could have been ordinary farm laborers. While this remains a distinct possibility we have presented arguments showing that the sharecropping hypothesis is nevertheless plausible and conforms well with the evidence. Another possible difficulty with the hypothesis arises from the fact that most type 5 entries, even in the four counties where they could have been sharecroppers, possessed some personal property and in a few cases fairly substantial amounts (see appendix F). At the same time, some number of type 2a tenants, the presumed renters, possessed no or very little personal property, although these cases constituted a noticeably smaller proportion of the type 2 group. A defining characteristic of sharecropping was the furnishing of livestock, farm implements, and seed by the landowner. By contrast, renters furnished their own supplies. Thus if an individual reported personal property in the census, some of it could have consisted of farm implements and livestock.[28] Some of the type 5 entries seem to have possessed enough personal property to more than cover the value of these items on their farms. The type 2 renters pose even more serious difficulties. Why did some of them report no personal property when their farms clearly had implements and livestock? First of all, the value of personal property, particularly in the lower range, can be defective. A great deal of confidence cannot be placed in its accuracy.[29] Second, the "rule" that landlords furnished croppers but not renters did not always hold, at least in the postbellum period, as Mangum has pointed out.

> Sometimes it is difficult to ascertain from a given set of facts just what relationship there was an intent to create. Border line situations are continually arising, and the cases are not always reconcilable. If the landlord supplies the land only, the courts will in all probability rule that a tenancy relationship is created; if he supplies everything except the actual labor, this will constitute a cropper contract. *Rarely does such a clear-cut instance arise, however,* for most of the cases are complicated by the fact that the landlord and tenant each supplies in whole or in part some of the other things which are needed for the effective use of the land, such as livestock, seed, fertilizer, tools, and farming machinery. Every case must be decided according to the intention of the parties as shown by the contract under consideration and by the facts surrounding its execution and performance.

Thus the possession of personal property, even if it included farm implements and livestock, cannot be taken as decisive evidence that an individual was a renter. By the same token, the absence of personal property did not necessarily imply cropping. Mangum has cited an 1885 Arkansas case in which the court ruled that the wording of a tenancy contract implied a rental agreement even though the landlord furnished *all* the supplies.[30]

Despite the fact that most type 5 farmers possessed some personal property, they were, especially in those counties where they could have been sharecroppers, substantially poorer as a group than tenants classified as type 2a. It is noteworthy that in most of the sample counties in which type 5 entries were most likely to have been mainly, or at least in part, first-year farmers, their personal property holdings were not much different from that of type 2a tenants. In a few counties such as these the type 5s were even wealthier. In Thomas County, in which type 5 could have comprised both first-year farmers as well as all sharecroppers, 52.7 percent reported no personal property at all, while 13.5 percent reported personal property equal to or in excess of $1,000.[31]

It has been shown that in at least four of the sample counties a case can be made that type 5 farmers were tenants, possibly sharecroppers. First, it is unlikely that they were merely first-year farmers who failed to meet the production criteria for entry on schedule IV, because the enumerators in the four counties, in violation of their instructions, apparently did enter first-year farmers on that schedule, where many of them appear as types b and c. Second, the use of the standard convention for type 2 tenants and the failure to report output for type 5 farmers on schedule IV in the same counties make sense in terms of legal distinctions between renting and cropping. Third, although the potential croppers generally reported some personal property, they were as a group substantially poorer than the potential renters. A difficulty with the hypothesis is that three of the counties showed no farm laborers. But other heads of household in these counties were designated "laborer," "day laborer," or "hireling." Some of them could have been ordinary farm laborers. In another county, Echols, most of the designated "laborers" must have been engaged in agriculture and may have been tenants of some kind and quite plausibly sharecroppers, since renters are separately designated on the returns. Clayton County offers the clearest evidence that the term *farm laborer* could be applied to tenants generally but in this case without distinguishing between forms of tenancy.

Obviously enumerators had very different perceptions of the social relations that prevailed among the free agricultural population of antebellum Georgia. To the extent that *farmer* implied membership in a particular social category, some enumerators, like the one in Clayton, assumed that tenants did not qualify. Others, like the Echols enumerator, may have placed renters in that exalted group but not sharecroppers, or perhaps share renters. Still others, the enumerators in Habersham, Harris, Houston, and Twiggs, were quite generous in allowing all sorts of folk to enter the farmer's domain: proprietors, renters, and possibly even sharecroppers or at least farm laborers.

It is not surprising that enumerators were frequently inconsistent among themselves in the mode of reporting tenancies; their instructions were virtually silent on the matter. Even after the Civil War, confusion abounded over the legal ramifications of different types of tenancies. Some antebellum enumerators seem to have made no attempt to distinguish tenancy types. In Worth County, for example, type 5 entries were almost certainly first-year farmers who were tenants rather than a distinctive kind of tenant, and all type 2 entries reported improved acres and farm value (and in almost all cases unimproved acres as well). Types 4 and 5 were most likely first-year farmers in Worth, since they were relatively few and the county reported no types b and c. Type 2 entries in Worth were designated "tenant" on schedule IV. Whatever forms of tenancy may have existed in Worth, the enumerator simply assigned all acreage and farm value to whoever operated the farm. In Towns County, on the other hand, in which there were only six type 5 entries, the enumerator followed the standard convention uniformly for the 172 type 2 entries. It may be that the Towns enumerator was following the same practice as his colleagues in the four counties discussed in the preceding paragraphs and that there may have been a handful of sharecroppers, appearing as type 5, in the county. More likely, he, like the enumerator in Worth, used only one convention for all types of tenants, but in this case it was standard rather than full values.

Although the Towns enumerator uniformly adopted the standard convention for tenants, Fannin and Gilmer, two other mountain counties examined by the short method, revealed each in turn very different enumerator practices. In Fannin the enumerator entered a small number of tenants in standard convention on schedule IV (13 out of 327 total entries) whose occupations were variously given as "farm tenant," "farmer," or "farm hand" on schedule I. On the latter schedule he reported 307 farm hand household heads out of a total of 906 households. Most of these farm hands, as far as we can judge from the short method, were not entered on schedule IV, but they must nevertheless have been tenants of some kind. If not, Fannin would have had a very different sort of social structure than neighboring Towns, where the type 2 tenancy rate was 40.6 percent. The Gilmer enumerator was also unique. He entered on schedule IV a large number of tenants reporting both improved acres and farm value and a smaller but still significant number reporting unimproved acres as well.[32] The different modes of reporting, and even apparent internal inconsistencies, in these three mountain counties obviously make any generalizations about forms of tenancy in that region highly speculative at best.

The same convention in reporting acreage and farm value for tenants did not necessarily mean the same thing to different enumerators. Some enumerators employed the standard convention for all types of tenants while others reported, for example, only improved acres for all types of tenants, with neither group of enumerators distinguishing type 5 or farm laborers as a separate category. But in many counties the same enumerator regularly employed two or more conventions on schedule IV, and unless he was simply being inconsistent—and the substantial number of counties in which this practice occurred in large numbers of records argues against mere inconsistency—he could have had some real distinctions in mind. Appendix A reveals the multiplicity of practices in the counties examined by the short method. A comparison of this table with appendix B will also allow one to compare the use of schedule IV conventions with the presence of type 5 and type 3 entries, the latter possibly constituting, in effect, a separate schedule IV convention. Type 5, particularly in counties with few or no types b and c, may itself be a convention for identifying sharecroppers. In counties where three or more conventions appear to some significant extent, the enumerator may have been making finer distinctions between the numerous types of tenancy arrangements that could occur. Whether these arrangements included sharecropping or simply various types of rentals is impossible to say on the basis of the census returns alone.[33] Since historians have, however, adduced other evidence for a variety of rental arrangements, it is not implausible that at the very least multiple conventions in a single county reflected such arrangements.

In addition to using a variety of conventions in recording acreage and farm value for tenants, some enumerators explicitly identified tenants as such, usually in the margin, name column, or the acreage–farm value columns of schedule IV, but sometimes in the real property or occupation column of schedule I. In some counties, Baldwin, for example, virtually all tenants were recorded in a single convention and identified as such, in this case with improved acres only and by "rented" written in the unimproved and farm value columns. But in Jasper County, although all tenants but one were recorded in standard convention, according to short-method estimates, only about two-fifths of them were identified by "tenant" written in the improved acreage column. Other similar examples could be cited. Moreover, in some counties the pattern could become complex. In Elbert, for example, fewer than half the standard convention entries were identified on schedule IV as "rented" or "rented land," while all but three of the tenants with improved acres only were so identified.

Further evidence for multiple forms of tenancy, including sharecropping, comes from a small number of counties each of whose enumerators applied explicit but differing designations for tenants, in all cases in the occupation column of schedule I. Table 5.3 presents these variants for Madison, Paulding, and Walton counties. Schedule IV conventions for these counties, as estimated by the short method, can be found in appendix A. First of all, the three enumerators all used the designation "cropper" (or "croper" in the case of Madison) in the occupation column of schedule I. Whatever its actual legal meaning may have been, there can be no doubt that the term itself was in current use in Georgia in 1860.[34] Moreover, in Madison, "tenants," and in Paulding both "tenants" and "renters," were separately distinguished. Madison, too, contained a very large number of "farmers" on schedule I who were heads of household with no real property, while Paulding reported an unusually large number of "day laborer" heads of household on that schedule. These counties, then, on the basis of real property holdings and occupational designations, included candidates for several different kinds of tenancy.[35]

But the modes of reporting in the three counties raise more questions than can be answered on the basis of data currently available. Considering now schedule IV tenancy conventions, we can see that in Madison there is a rough equivalence between the number of tenancy candidates on schedule I and the number of schedule IV type a en-

Table 5.3

AGRICULTURAL AND LABORING OCCUPATIONS DESIGNATED
ON SCHEDULE I IN THREE COUNTIES, 1860

	Madison	Paulding	Walton
Heads of Households—Without Real Property			
Farmer	144	11	177
Day Laborer	38[a]	113	65[b]
Farm Laborer	1	5	1
Tenant	8	210	—
Renter	—	14	—
Cropper	8	14	20
Heads of Households—With Real Property			
Tenant	1	19	—
Renter	—	4	—
Cropper	—	1	1
Non-Heads of Households—Without Real Property			
Tenant	2	8	—
Cropper	2	3	71
Non-Heads of Households—With Real Property			
Cropper	—	—	1

[a]Includes thirty-six entries designated "laborer."
[b]Includes one entry designated "hireling."

tries reported in the two conventions, although the sixteen nonstandard entries were probably not tenants.[36] Clearly the distinctions in occupational designations were not reflected in different tenancy conventions on schedule IV. Our examination of sample pages for the short method turned up "farmers," "croppers," "laborers," and "day laborers," entered in standard convention.[37] In contrast to Madison, the majority of schedule I tenancy candidates in Paulding did not appear in tenancy conventions on schedule IV, although some number were reported with full values on the latter schedule and would appear as type 3.[38] But even taking type 3 into account almost certainly leaves the majority of tenants unreported by convention on schedule IV. Moreover, we found examples on schedule I of type 5 "farmers," "tenants," "renters," and "croppers" whom we could not locate on schedule IV by the short method. While the great majority of entries in both standard convention and with improved acres only were identified as "tenant" on schedule I, one of the four type 3 records found by the short method appeared as a "cropper" on that schedule. The enumerator in Walton County employed both a triple convention on schedule IV and identified three possible groups of tenancy candidates on schedule I: "farmers," "day laborers," and "croppers"; but we were able to locate no croppers on schedule IV, and even the number of "farmers" on schedule I who were heads of household and reported no real property is substantially less than the number of entries in all tenancy conventions on schedule IV, implying the existence of type 5 entries who could have been first-year farmers or a distinctive kind of tenant.[39]

The apparent internal inconsistencies in Madison, Paulding, and Walton counties are not readily explicable. The enumerators may have been confused by the lack of clear instructions from the census office or they may have been altogether whimsical in their choice of procedures. It is perhaps not entirely coincidental that all three counties are located in the upper Piedmont. This region was a zone of transition on the upper margins of the cotton belt and may have been characterized by complexities in tenancy arrangements that were less likely to be encountered in more homogeneous areas. Steven Hahn has provided evidence of considerable variety among rental contracts in the upper Piedmont.[40] The use of the term *cropper* by the three enumerators at least suggests some sharecropping. In any case these enumerators were more or less familiar with a variety of terms and conventions. Inconsistency, confusion, and even whimsy usually have a basis in reality and, if so, may have expressed in some fashion the complex social and economic relationships that actually existed in antebellum Georgia agriculture.

We know that in the postbellum period, in addition to the fundamental legal distinction between renters and sharecroppers, actual tenancy arrangements varied widely with respect to modes of payment (cash, share of output, specified quantities of output), the degree to which the tenant or proprietor furnished supplies, and the division of responsibility between tenant and landlord regarding crops to be planted, tasks undertaken on the farm, and day-to-day managerial decisions. Other historians have found direct evidence of various kinds of tenancy contracts in the antebellum South, including Georgia, but in some cases have denied that the institution of sharecropping, properly so-called, existed to any significant extent. This is a problem of fundamental importance. If sharecropping was significantly present during the antebellum period, serious doubts would be raised about the hypothesis, as it is presently formulated, that proletarianization in southern agriculture, associated with the rise of sharecropping, began only after the Civil War.

Our examination of the census data has been unable fully to resolve the knotty problems surrounding forms of tenancy in antebellum Georgia. We have, however, attempted to establish some minimal patterns of intelligibility amidst the seeming chaos of different enumerator procedures. Too often historians have simply ignored their great variety and have not tried to make sense of apparent inconsistencies in modes of reporting both by a single enumerator and among enumerators. We have ventured into the chaos partly to show that the data can plausibly be interpreted in such a way that is consistent with the existence of a multiplicity of tenure forms.

We may conclude, first of all, that agricultural organization in antebellum Georgia appears more complex than historians have generally recognized. Any simple dichotomy between proprietor and tenant, or proprietor and farm laborer, seems unlikely. There were marked differences among enumerators with respect to their recognition of these complexities—differences that doubtless often had much to do with confusion about their nature. Some enumerators did make only a simple distinction between tenant and proprietor; but others may have attempted to deal with complexities by means of various schedule IV tenancy conventions, the decision to include some tenants on schedule IV while excluding other tenants, and the use of a variety of occupational designations for tenants, such as "farmer," "farm laborer," "tenant," "renter," and "cropper."

Second, some circumstantial evidence points to the existence of sharecropping in antebellum Georgia, though perhaps not in all regions. A few enumerators did use the term *cropper*, although they could have been applying it loosely to share renters. Additional evidence for the existence of sharecropping comes from the decision of certain enumerators to exclude some apparent tenants from schedule IV altogether. Only if the enumerator reasoned that the tenant was not in possession of the crops would such exclusion have made sense. Usually these excluded individuals were called farmers, but the use of the term *laborer* in Echols County, for example, may have reflected even more sharply the legal distinction between renter and cropper.

We must emphasize again that our conclusions, particularly about sharecropping, are of a highly provisional nature. We can hope that a series of fully comparable microstudies, extending the scope of our preliminary research to other states and regions and through time, and relying on additional data in county archives and plantation records, will provide some of the answers to questions we have raised. But we may have to reconcile ourselves to the possibility that much of the data required for such an undertaking simply do not exist.

One question of importance remains. If it is true that enumerators sometimes excluded some tenants from schedule IV and identified them only on schedule I as farmers not reporting real property (type 5) or simply as farm laborers, how can a reliable estimate of tenancy rates, comparable across all counties and between census years, be made? We have approached—but not fully solved—this problem by presenting tenancy estimates in chapter 6 at four levels. Level I counts only type 2 entries as tenants. These certain tenants are entered in a tenancy convention on schedule IV and report no real property on schedule I. Their number represents the irreducible minimum of tenants in any county. The reliability of the count is attested by the fact that some enumerators explicitly identified entries, otherwise classified as type 2, as tenants.[41] Level II adds type 3 entries to the tenancy count. Such entries also appear on both schedules and report no real property on schedule I. Since type 3 operators are not entered in a convention but, like owner-operators (type 1), report acreage and farm value, they could be errors for whom the enumerator failed to record real property. But the reporting of acreage and farm value might itself constitute a tenancy convention. Moreover, type 3 records could be cases of exclusively tenanted farms which were assigned values in lieu of an absent proprietor. Finally, the size of type 3 holdings in most counties and the value of their production more nearly resemble those of other tenants rather than owner-operators (see appendix G). All these considerations lead us to believe that most type 3 entries were tenants, but since we are not certain, we did not include them in the level I estimate.

More problematic is the status of those farmers without farms (type 5) whom we have counted as tenants at level III. We have speculated that type 5 farmers may have been sharecroppers or tenants of a kind that some enumerators chose not to enter on schedule IV. In many of our sample counties the bulk of type 5 entries were probably first-year tenant farmers who did not yet show sufficient production to merit inclusion on schedule IV. Tenancy estimates that do not count such first-year farmers in 1860 will not be comparable to the tenancy rates obtainable from the published aggregate census in 1880. In 1860, farmers who had an output of less than $100 were supposed to be (but were not always) excluded from schedule IV. In 1880, however, the instructions stipulated that only farms of less than 3 acres, unless they sold at least $500 in produce, were to be excluded. Therefore, first-year farmers, unless they happened to hold less than 3 acres, would have been counted that year.[42] Clearly, some adjustments in the 1860 data, taking into account first-year farmers, need to be made in order to achieve comparability with the published tenancy rates for 1880. To the extent that type 5 may include sharecroppers, the value of a tenancy rate at level III is further enhanced. To the extent that type 5 entries may not have been tenants at all, but perhaps ordinary farm laborers, the reliability of the level III tenancy estimate is reduced.

Finally, we have calculated a tenancy estimate at level IV that adds farm laborers on schedule I only to the previous entry types. The status of farm laborers is clearly the most difficult to interpret, but we decided to include them at level IV because of the compelling evidence that in some counties enumerators almost certainly gave tenants a laboring occupation. We have shown, for example, that in Clayton County the choice is between assuming a virtual absence of tenants—and this is most implausible for the upper Piedmont region—and more plausibly assuming that farm laborers on schedule I only were actually tenants. We have also shown that the laborers in Echols County could easily have been tenants. For most counties we have no obvious way of knowing whether a farm laborer was just that or some kind of tenant.

The estimation of tenancy rates at four levels is by no means an altogether satisfactory solution to the problem of identifying tenants in the census returns. At level III and particularly at level IV we make no claim that the estimates are reliable for any particular county or that they are necessarily appropriate for regions outside of Georgia. The purpose of the four levels of estimation is to suggest a reasonable range in which actual tenancy rates occurred. The derivation of the tenancy levels will be discussed at greater length in chapter 6. There we will also attempt to show which levels of estimation make the most sense in terms of the spatial distribution of tenancy in Georgia as a whole.

6

The Spatial Distribution of Tenancy in 1860

THE DISCUSSION IN THE preceding chapter makes it appear plausible that some large proportion of type 5 records and "farm laborers" were in many counties tenants of some sort, and possibly croppers; but it is important that these conclusions not be considered definitive. The case is a circumstantial one at best. Its strength might be enhanced, however, if it could be shown that the inclusion of type 5 entries and of farm laborers produces patterns of tenancy distribution that appear more rational, that is, accord more satisfyingly with the known spatial distributions of other variables. We shall therefore proceed to construct estimates of tenancy rates at several levels, each level incorporating a different set of record types into the numerator and denominator. Minimum and maximum tenancy estimates will thereby be provided against which various hypotheses about tenancy can be tested. But before these hypotheses are tested by the various levels of estimation, we propose to test the levels themselves by examining the relative plausibility of each when compared with a set of hypotheses that we consider, on purely logical grounds, to offer the best explanation of tenancy distributions.

Such a procedure may seem curious and even startling to those who are accustomed only to testing hypotheses with data, since we are suggesting that on some occasions the procedure might usefully be reversed by testing data against plausible hypotheses. The former procedure is effective as an initial procedure when the data are well or sufficiently understood, as is the case with much modern economic data. But historical data present special problems that cannot be overlooked. Such data are characteristically not well understood. Our procedure seeks to understand data by searching for *patterns of intelligibility*. Some amount of anomaly must be expected, of course, but pattern should be evident. In the present study, we expect that counties that share a geographical locality and that are in every known respect economically, geophysically and demographically generally comparable will not vary sharply in the structure of their agrarian regimes or, more pointedly, in their tenancy rates. We look for patterns of regional, or subregional, homogeneity. Such expectations constitute working assumptions; to adopt contrary assumptions seems to us to undermine the very foundations of empirical investigation. Our first requirement, then, is that regionally distinct patterns emerge. If they do not emerge, either the data are fatally flawed, or are insufficiently understood, or the world being examined is a very different sort of place than we imagined. Second, if patterns emerge and can also be shown to conform to hypotheses that are plausible, either *a priori* or because of earlier work done in the field, then our confidence in both the intrinsic reliability of the data and our understanding of them is greatly enhanced. In the following pages we shall see that patterns of intelligibility for tenancy displaying regional and subregional characteristics do emerge but that the patterns vary depending on how the county rates are computed. Choices of which mode should be preferred depend wholly on how the data should be understood. Our preference of mode will be dictated by how the regional patterns cohere internally and by how well such patterns conform to hypotheses that have a higher prior probability. The hypotheses are not essential to the test, and the patterns that emerge may immediately suggest alternatives; but an analysis of the patterns in the context of such hypotheses can lend credibility to the inferred meaning of the data. The process is essentially retroductive. Once the community of scholars working closely with historical data series can move toward a consensus on what such data mean, the process may reverse itself, and we shall be in a position to use such data for testing hypotheses in the normal manner.

It is possible to derive four distinct levels of estimated tenancy rates. The rates at each of the four levels in the sixteen sample counties, and the formulae by which they are derived, are displayed in table 6.1. Level I must be accepted as providing indisputable minimum rates of tenancy.[1] The only way to reject these rates as minimums is to reject type 2a records as tenants. But to reject them as tenants entails reject-

Table 6.1

FOUR LEVELS OF ESTIMATED TENANCY RATES (%), SIXTEEN SAMPLE COUNTIES, 1860

County	Level I[a]	Level II[b]	Level III[c]	Level IV[d]
Baker	17.0	20.3	29.2	33.3[e]
Baldwin	18.1	25.5	25.1	26.2[e]
Clay	18.1	19.6	23.9	26.3[e]
Coffee	27.3	28.5	29.6[e]	29.6
Echols	20.7	23.1	23.1	46.2[e]
Emanuel	13.2	14.4	17.3[e]	19.0
Greene	7.6	9.0	16.7	17.2[e]
Gwinnett	20.1	21.0	27.8	32.6[e]
Habersham	37.1	37.6	41.1	41.6[e]
Harris	19.1	20.5	26.3[e]	27.6
Houston	10.6	10.6	18.7	19.6[e]
Thomas	9.2	13.7	34.6	36.0[e]
Towns	40.6	41.9	42.6[e]	42.6
Twiggs	31.3	31.5	34.9[e]	37.0
Wilkes	3.4	10.7	14.2[e]	26.3
Worth	18.2	22.0	23.1	29.9[e]

[a] Level I = type 2a/types 1a + 2a + 3a + 6a + 7a
[b] Level II = types 2a + 3a/types 1a + 2a + 3a + 6a + 7a
[c] Level III = types 2a + 3a + 5/types 1a + 2a + 3a + 4 + 5 + 6a + 7a
[d] Level IV = types 2a + 3a + 5 + farm laborers/types 1a + 2a + 3a + 4 + 5 + 6a + 7a + farm laborers
[e] Final preferred maximums

ing the clear evidence of their meaning found in counties where tenants are annotated as such. Of even greater consequence, their rejection requires an assumption that standard convention and most other entries we identify as tenancy conventions are merely defective. Such entries constitute so enormous a proportion of those found on schedule IV that, if this conclusion were reached, one must entirely abandon the census as so pervaded by error as to be worthless. So radical a conclusion is clearly unwarranted. The problem, then, is not to establish the fact of tenancy or its minimum rate so much as it is to establish a maximum range within which the actual rate must fall. Levels II through IV are designed to explore three alternative methods for establishing an upper limit to the range.

Level I estimates (map 6.1) produce a spatial pattern of rates that in most instances accords well with what one would have expected to find, or at least with one set of plausible assumptions. Tenancy finds its lowest incidence on the old and relatively exhausted lands of the eastern lower Piedmont, where the rate is well below 10 percent, while the fresher lands of the middle to western lower Piedmont and the Central Cotton Belt have rates approaching 20 percent. The rates in the mountains are the highest in the state, exceeding 40 percent in Towns, with a slight decline as one descends southward down the ridge into the hill country. Even the Pine Barrens show interesting rational differences, varying from 13 percent in Emanuel, which lies in one of the earliest settled sections of the state, to 27 percent on the fresher lands of Coffee. These variations are consistent with a thesis that tenants tended to move with patterns of settlement, seeking more productive lands in their own self-interest, while proprietors, especially in counties where absenteeism was prevalent, encouraged the entry of tenants into such regions as an alternative and supplementary labor supply to slavery. There is some evidence that tenants were placed on the poorer and intermediate lands in the newer regions of staple production, the owners applying their slave labor principally to the most productive soils. Tenancy in the cotton counties on the eve of the Civil War, then, may perhaps be related to clearance and to the flexible use of an alternative labor supply when the application of slave labor, the most capital intensive form of supply, seemed unwarranted or when the demand for slave labor on the new lands exceeded its supply. An explanation related to the appropriateness of slave labor is similarly applicable to the mountains and the Pine Barrens and even to the Palmetto Flats of southeastern Georgia. The relative lack of a staple economy in these regions effectively prevented slave labor's being drawn into them. A system of white tenancy was therefore the only solution to clearance and to the farming of any lands not directly under the control or within the immediate labor capacities of proprietors. We believe such hypotheses are the most logically plausible ones available for explaining tenancy distributions. Our discussion of the four levels of tenancy estimates will therefore proceed by testing the plausibility of each estimate level within the framework of these hypotheses.

The rates established at level I generally support the hypotheses, but they also pose serious interpretive problems. The differences in

Map 6.1. Tenancy Rates (%), Level I, Sixteen Sample Counties, 1860

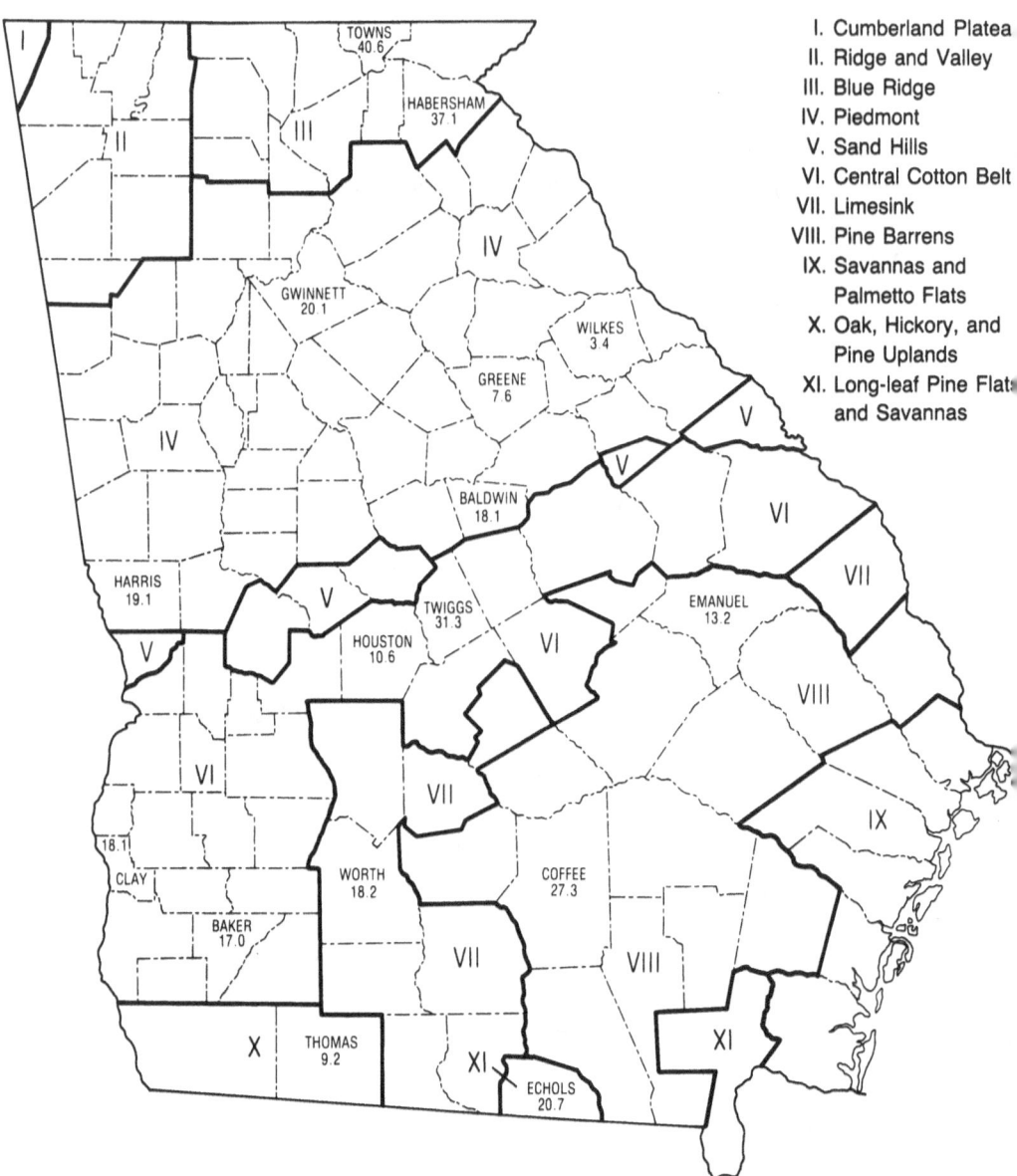

I. Cumberland Plateau
II. Ridge and Valley
III. Blue Ridge
IV. Piedmont
V. Sand Hills
VI. Central Cotton Belt
VII. Limesink
VIII. Pine Barrens
IX. Savannas and Palmetto Flats
X. Oak, Hickory, and Pine Uplands
XI. Long-leaf Pine Flats and Savannas

Notes: Boundaries of economic regions are indicated. For the data, see table 6.1.

rates between Houston and Twiggs is jarring. Judging from the 1880 report, the lands in Houston were among the best in the Central Cotton Belt, and one might therefore expect to find a higher use of tenants in Twiggs; but the magnitude of difference is too great. The problem lies not so much in the Houston values, which are below but not radically out of line with others in the southwestern region given the special qualities of its soils; the problem lies principally with the high values of Twiggs, which accord ill with values in all other counties heavily under staple production. Thomas County, on the other hand, has too low a rate. Granted its lands were not so favorable to cotton and did not provide incentives so favorable to clearance and production as those of the Central Cotton Belt, nevertheless Thomas was a substantial producer of cotton on soils of intermediate quality, enjoyed a thriving livestock economy on its uplands, and was still in 1860 undergoing considerable clearance. Even more striking are patterns of rates found in the southwestern cotton counties, Worth in the Limesink and Coffee in the Pine Barrens. There is no clear reason why Worth should display a rate precisely conformable to those under heavy staple production to its west. The sharp differences in black labor supply to be found in Worth and in the southwestern cotton counties may be seen on map 7.8. The 22.9 percent slave population in Worth was far lower than any counties to the west of it. Our expectation would be that tenancy rates would rise as one proceeded from the cotton counties into the Limesink, and again as one left the more productive lands of the Limesink and entered the largely uncleared and relatively fresh but highly acidic lands of the southwestern Pine Barrens. Black labor was relatively uneconomic on these poorer soils, and the production of staple sharply declined. The expectation at least is fulfilled in the transition from the Limesink to the Pine Barrens, but the magnitude of the transition is brought into doubt by the low reported value in Worth. Similarly, we expected a more marked transition as one entered the upper Piedmont counties where cotton production and black labor supplies drop off sharply. The rate for Gwinnett is therefore low, being almost identical to that of Harris.

The estimates at level II (map 6.2) remove some of these anomalies in spatial distribution. While several counties are virtually unaffected, a more suitable transition has emerged in the belt of counties running from the southwest into the Pine Barrens. A tighter relationship has also developed between Wilkes and Greene, two counties that should

Map 6.2. Tenancy Rates (%), Level II, Sixteen Sample Counties, 1860

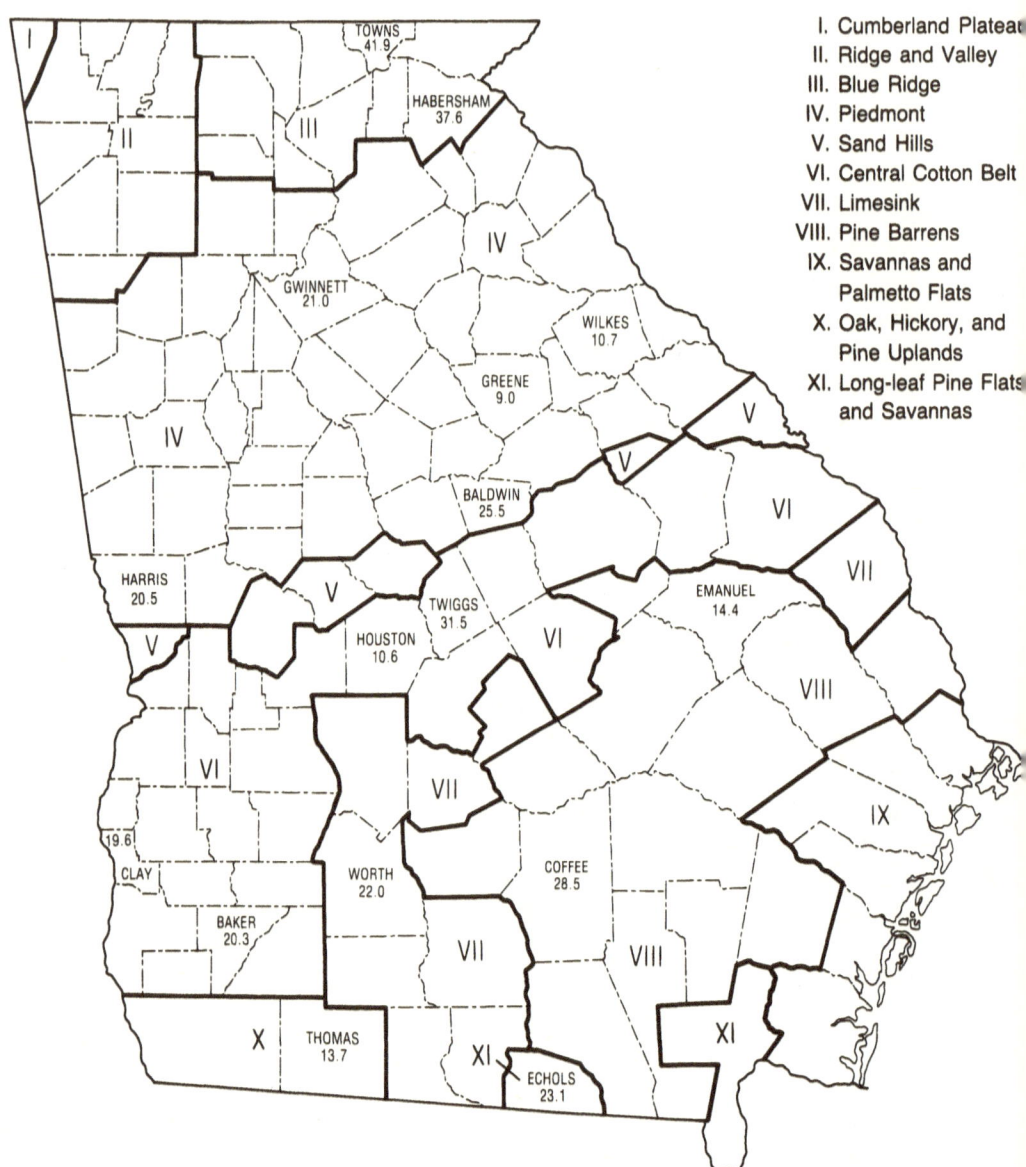

Notes: Boundaries of economic regions are indicated. For the data, see table 6.1.

in every respect be highly comparable. But Thomas is still lower than might be expected, the position of Twiggs within the Central Cotton Belt has in no way improved, and Gwinnett persists in behaving as though it were in the intensive cotton lands of the lower Piedmont. Moreover, as we have seen, a strong case can be made that types 2a and 3a do not exhaust all forms of tenancy to be found in antebellum Georgia. If the inclusion of additional types produce patterns of rates that conform more closely to rational expectations, given what we know of other variables, then that very rationality may be used to bolster our confidence in alternative rate estimates.

The estimates at level III (map 6.3), which take into account the "farmers without farms" (type 5), produce rate increases of considerable magnitude in many counties. The spatial effect is strikingly more satisfying in its conformity with other known variables. While Twiggs is still an upward outlier within the Central Cotton Belt, it now bears a more reasonable relationship to the newer cotton counties of the southwest. Within the southwestern corner a more rational transition has emerged as one proceeds southward out of the prime cotton lands into the more mixed economy of Thomas. The belt extending from Clay to Coffee fails to display a terrace as it enters the Limesink, but the relationship between Limesink and Pine Barrens, and within the Pine Barrens, maintains itself satisfactorily. Wilkes and Greene continue in tandem as they ought. The mountains never change. Gwinnett remains unexpectedly low, however.

The estimates at level IV (map 6.4), which further incorporate farm laborers into the computation, reveal a pattern in which most anomalies are sufficiently resolved to support the hypotheses associating tenancy with the quality of soil, the structure of the agrarian regime, and the allocation of labor for clearance and production. The lower Piedmont has a suitable transition from old to new lands, and Gwinnett is at last ideally placed between the intensive cotton lands of Harris and Baldwin and the Blue Ridge counties lying outside the staple economy. Twiggs is now sufficiently close to the rates of the southwest that its higher value may be attributed to those of its militia districts that lie in the Sand Hills, a region that contains significantly higher rates of tenancy than do the cotton districts on its northern and southern boundaries. A plausible transition is also maintained between Clay and Thomas. The Palmetto Flats of Echols now take on tenancy rates similar to those found in the Blue Ridge, alleviating a problem

Map 6.3. Tenancy Rates (%), Level III, Sixteen Sample Counties, 1860

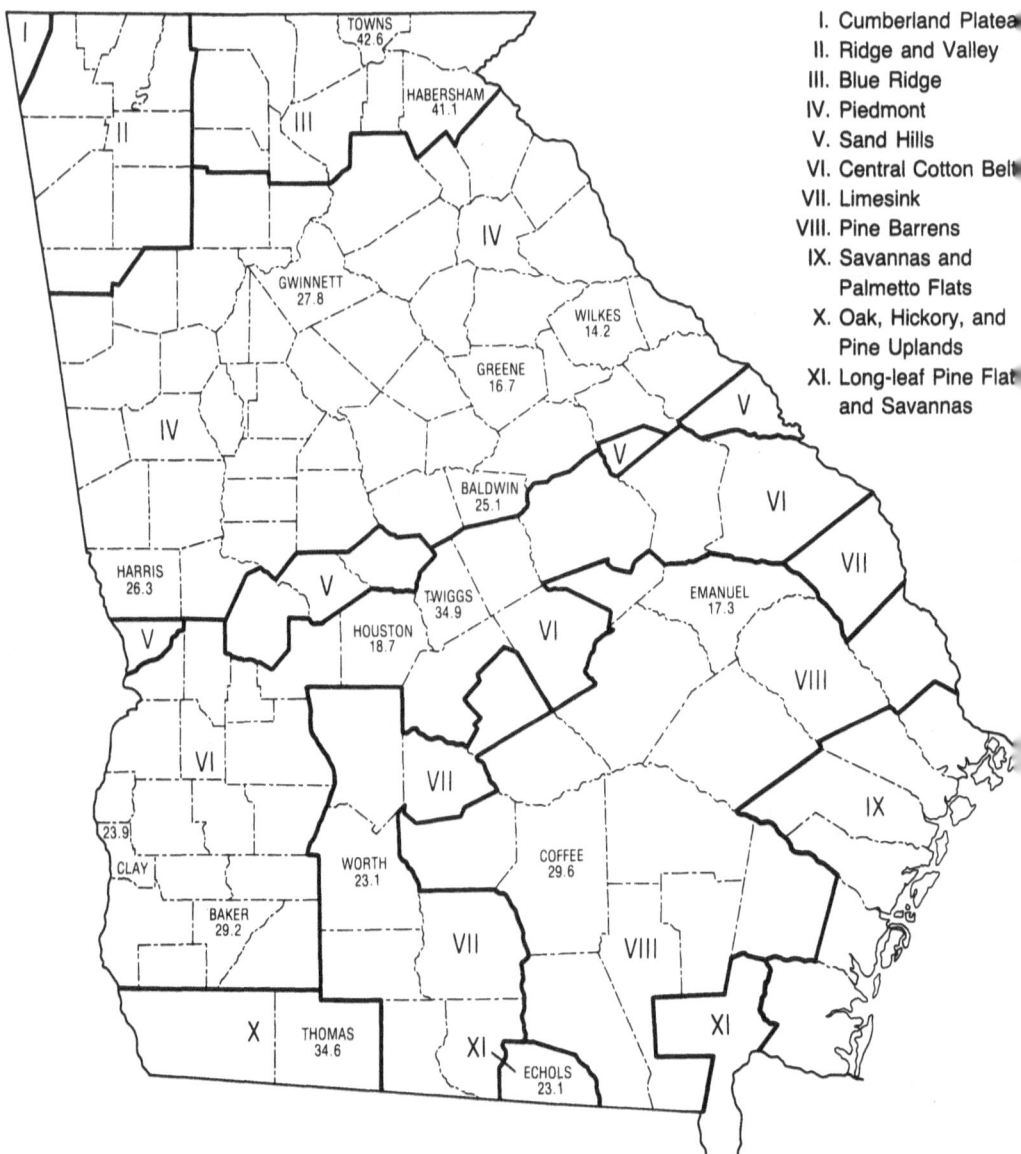

Notes: Boundaries of economic regions are indicated. For the data, see table 6.1.

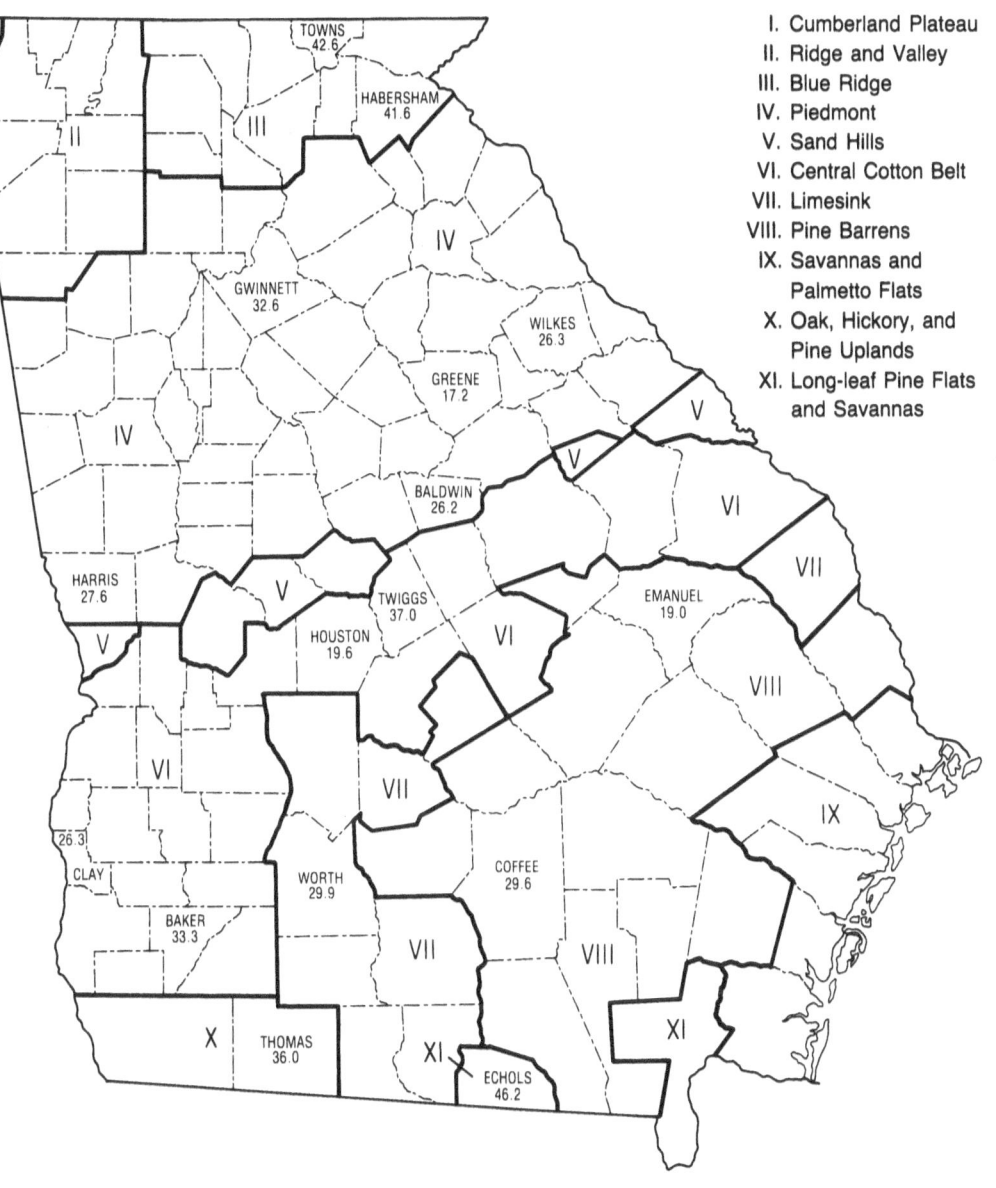

Map 6.4. Tenancy Rates (%), Level IV, Sixteen Sample Counties, 1860

Notes: Boundaries of economic regions are indicated. For the data, see table 6.1.

that is otherwise difficult to explain: why there should be radical differences in tenancy rates between the two spatially marginal areas of the state, both strongly characterized by the poorness of their soils and their remoteness from the cotton economy. It should be noted, however, that the form of reporting that brings Echols into line with the Blue Ridge is quite different from that encountered in the mountains. Tenants in Towns are reported in standard convention, with almost no laborers, superficially suggesting renters; while the higher rates in Echols depend in part upon tenants identified as such on schedule I and reporting both improved acres and farm value on schedule IV, and on a large number of farm "laborers" reported on schedule I only. The pattern in Echols thus suggests a mix of different forms of tenancy, possible renting and sharecropping.[2] Not all problems in the southern regions are resolved, however. Worth still seems at least marginally low with respect to the adjacent cotton counties, and quite out of line when one considers its strikingly lower slave population. On the other hand, slaves rather surprisingly formed an almost identical proportion of the population in both Worth and Coffee, which would make identical tenancy rates more credible. It may be that Worth, being adjacent to prime cotton lands, was subject to labor being drawn off onto the better lands of its western neighbors. It may also have had a high proportion of small proprietors who were drawn to its soils but lacked the capital and incentives to use high levels of slave or tenant labor. The region is a difficult one to analyze. The transition within the Pine Barrens from the older lands of Emanuel to the fresher lands of Coffee and on to the Palmetto Flats of Echols is at least in the expected upward direction.

All four levels of estimation conform reasonably well to the proposed hypotheses. But the tightest conformity emerges at level IV. This result does not by any means compel one to conclude that type 5 records and farm laborers represented tenants of some sort in any particular county, much less in all counties. Alternative explanations may be offered for the success of level IV. For example, it might be argued that level IV is so enlarged that it includes virtually all heads of household engaged directly in the rural economy. Its spatial patterns may therefore merely reflect distributions of labor and capital in a more general form rather than tenancy in particular. Such an observation cannot easily be faulted, since the argument for tenancy is a subset of such an argument. The argument for tenancy must therefore be based on further specifications that narrow the range of plausible alternatives. Such specifications were provided by the arguments of the preceding chapter, examining the alternative interpretations of such record types and the meaning imposed on tenancy and sharecropping by the courts and attributed to them by persons familiar with landholding relations in the South during the late nineteenth and early twentieth centuries. The combined weight of these arguments, while not conclusive, is highly persuasive; and while investigators are always well advised to understate rather than overstate their case, we feel we would be ignoring the preponderance of the evidence if we were to place the maximum estimation of tenancy rates for most counties at less than level IV. In accepting level IV as the maximum, however, it must be borne in mind that inconsistency of reporting among enumerators, which has been so central a theme in this study, makes it inevitable that level IV statistics will themselves be inconsistently reliable and lack comparability between counties to some degree. Some number of type 5 entries in most counties are undoubtedly first-year farmers, especially in those counties where types b and c are not reported and type 4 entries are substantially present. Type 5 entries may also simply be retired farmers without real property, younger farmers with no holding, or ordinary farm laborers. Some type 4 farmers may have no agricultural holdings within their county of residence. Even more probably, some number of designated farm laborers may simply be what the term normally connotes and hold no tenancy either as renter or sharecropper. In many counties, especially those where multiple conventions are employed, an investigator may decide that farm laborers should play no role in computing tenancy. The application of the procedures implied by level IV must at all times be undertaken with judgment and discretion and fully in the context of the enumeration techniques employed within each county and the pattern of tenancy rates found within the region.

Our final estimation of maximum tenancy rates for the sixteen sample counties is therefore a mixture of levels III and IV (table 6.1). Coffee and Towns are retained at level III for the simple reason that they reported no laborers of any description as heads of household. While Echols reported only "laborers" and failed to differentiate them into types, we have concluded that so large a number of laborers in this remote and exclusively rural county could only be farm laborers. We have therefore classified Echols at level IV. It is true that such a classification inflates our case by enhancing the spatial pattern that we favor, but alternative explanations are simply less convincing and seem implausible. Harris also reported only "laborers" and failed to differentiate them into types. In this case, since Harris is not a remote county and does display significant urban characteristics scattered throughout its various districts, it is not implausible that the sixteen laborers found there were unconnected with farming. We have therefore retained Harris at level III. In Wilkes County virtually all overseers, identifiable as such by annotations in the margin of schedule IV, were described as "farm laborers" on schedule I. Wilkes may be unique in this respect among the sixteen sample counties. All cotton counties reported substantial numbers of "overseers" on schedule I. Even outside the cotton counties, overseers were commonly encountered, though in smaller numbers. Only Coffee, Emanuel, and Towns reported none. We have therefore concluded that some large number, and perhaps all, "farm laborers" in Wilkes were in fact overseers, and we have classified the county at level III. It may be that an identical problem arises in Emanuel, where twelve "farm hands" are the only laborers reported only on schedule I and no "overseers" are reported. As a marginal cotton county, Emanuel should have reported some small number of overseers. Other counties with its characteristics certainly did so. We have therefore classified Emanuel at level III. Twiggs is classified at level III because its fifteen laborers were explicitly described as "day laborers," a description that is compatible with farm labor but seems entirely inappropriate for all varieties of tenant. All the remaining counties in the sample have been classified at level IV; with the exception of Echols, all counties classified at level IV differentiate between agricultural and other forms of laborers, and only those designated "farm laborer" or "farm hand" or—in the seven cases in Habersham and two out of twenty-nine cases in Worth—"hireling" have been included in the tenancy computation. The result, as displayed in map 6.5, forms a pattern that strikingly conforms to the hypotheses we have offered.

It is now appropriate to determine the reliability of estimated tenancy rates produced by the short method, which we have calculated for every class 1 and class 2 county in the state. We may begin by examining table 6.2, which gives tenancy rates derived by the long method at levels I and either III or IV and compares them to estimates derived by the short method in the sixteen sample counties. Two questions are pertinent: first, do the short-method estimates fall

Map 6.5. Final Estimates of Maximum Tenancy Rates, Levels III and IV, Sixteen Sample Counties, 1860

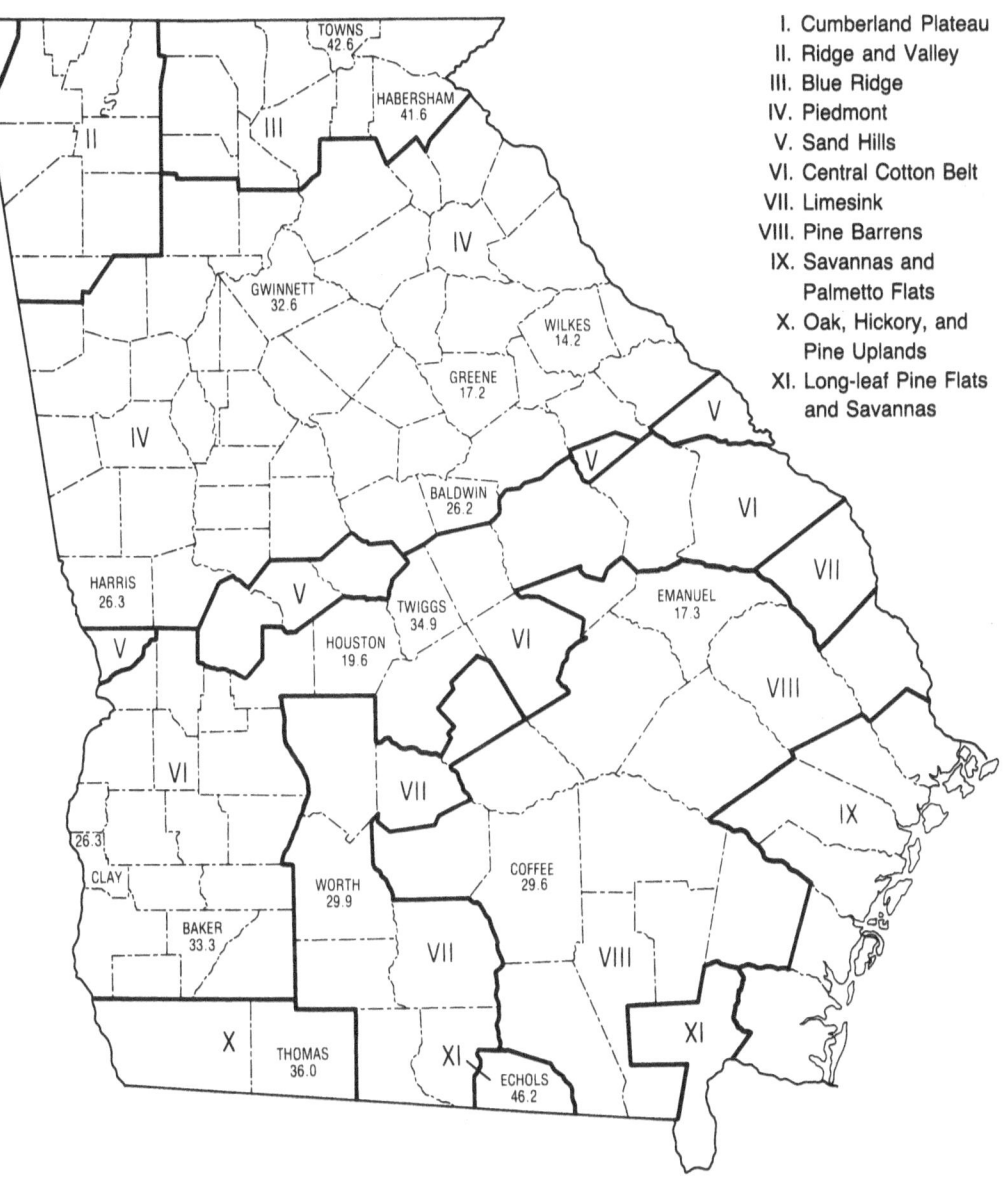

Notes: Boundaries of economic regions are indicated. For the data, see table 6.1.

Table 6.2

COMPARISON OF ESTIMATED TENANCY RATES (%)
BY THE LONG METHOD (LEVELS I AND III/IV) AND SHORT METHOD,
SIXTEEN SAMPLE COUNTIES, 1860

County	Short Method[a]	Level I	Level III/IV
Baker	27.7−	17.0	33.3
Baldwin	21.0	18.1	26.2
Clay	19.2	18.1	26.3
Coffee	29.4	27.3	29.6
Echols	31.4	20.7	46.2
Emanuel	14.3+	13.2	17.3
Greene	8.5	7.6	17.2
Gwinnett	22.1	20.1	32.6
Habersham	37.8	37.1	41.6
Harris	22.8	19.1	26.3
Houston	Class 3	10.6	19.6
Thomas	Class 3	9.2	36.0
Towns	43.2	40.6	42.6
Twiggs	35.4	31.3	34.9
Wilkes	3.4+	3.4	14.2
Worth	21.0	18.2	29.9

[a]Class 1 rates are underscored. For a definition of short-method classes, see chapter 2.

within the range of estimation established by levels I and III/IV; and, second, do the short-method estimates produce a pattern of spatial distributions that is compatible with the pattern of the intrinsically more reliable estimates of the long method? In answering the first question it is important to remember that the short method has not been designed to produce estimates that are accurate in the magnitude of their individual values; the method is in fact inherently incapable of such precision, and any precisely accurate values could be achieved only accidentally. The short method was designed to test for roughly accurate patterns of regional distribution and to detect conventions, and to do both only in a preliminary fashion.

An examination of table 6.2 shows the short-method estimates to be satisfactory in the magnitudes of their values. Every value except two, those for Towns and Twiggs, which are marginally high, falls within the range of minimum and maximum estimates established by the long method. The variations in conformity to levels I and III/IV are essentially due to variations in modes of reporting by different enumerators. But while such variation may always be expected, the results presented here indicate that the magnitude of short-method values may be expected to fall within the minimum and maximum range of levels I and III/IV. If this conclusion is accepted, then it follows that short-method estimates should produce a spatial pattern roughly in conformity with what might be produced by long-method estimates. The pattern will unavoidably be eccentric and erratic in its details, however, owing to the tendency of short-method values to vary within the range. The most important element influencing erratic variation will be the presence of type 5 records and farm laborers, when such records are present in substantial numbers. These records are not initially included in short-method estimates, and their presence may bias such estimates downward. Large numbers of type 3 records are similarly important but comparatively rare. An examination of map 6.6 will confirm that the short-method estimates for the sixteen sample counties tend to produce a pattern of distributions that conform rather closely to those derived by the long method. The only serious difficulties lie in the high value for Twiggs when compared with other Central Cotton Belt counties and in the inability to produce sufficiently high values in Gwinnett to detect an upward swing as one enters the upper Piedmont. Estimations derived by the short method will not necessarily reveal all major variations in spatial distribution; and, more particularly, variations within regions and the specific values for individual counties may be expected to be erratic and unreliable. But, with these important reservations clearly in mind, estimations of tenancy rates derived by the short method seem to produce patterns of distribution that serve well as a first approximation of regional differences.

Map 6.6 presents our estimates of tenancy rates derived by the short method for each county in Georgia. By comparing the map with regional tenancy patterns suggested by the long method, and by noting whether short-method patterns conform reasonably well to plausible regional hypotheses, we may simultaneously learn more about the reliability of short-method estimates and enhance our understanding of tenancy patterns within Georgia. It will be seen that not every county was amenable to analysis within classes 1 and 2,[3] but we were able to derive values for a number of counties sufficient for anal-

Map 6.6. Tenancy Rates (%), Short-Method Estimates, 1860

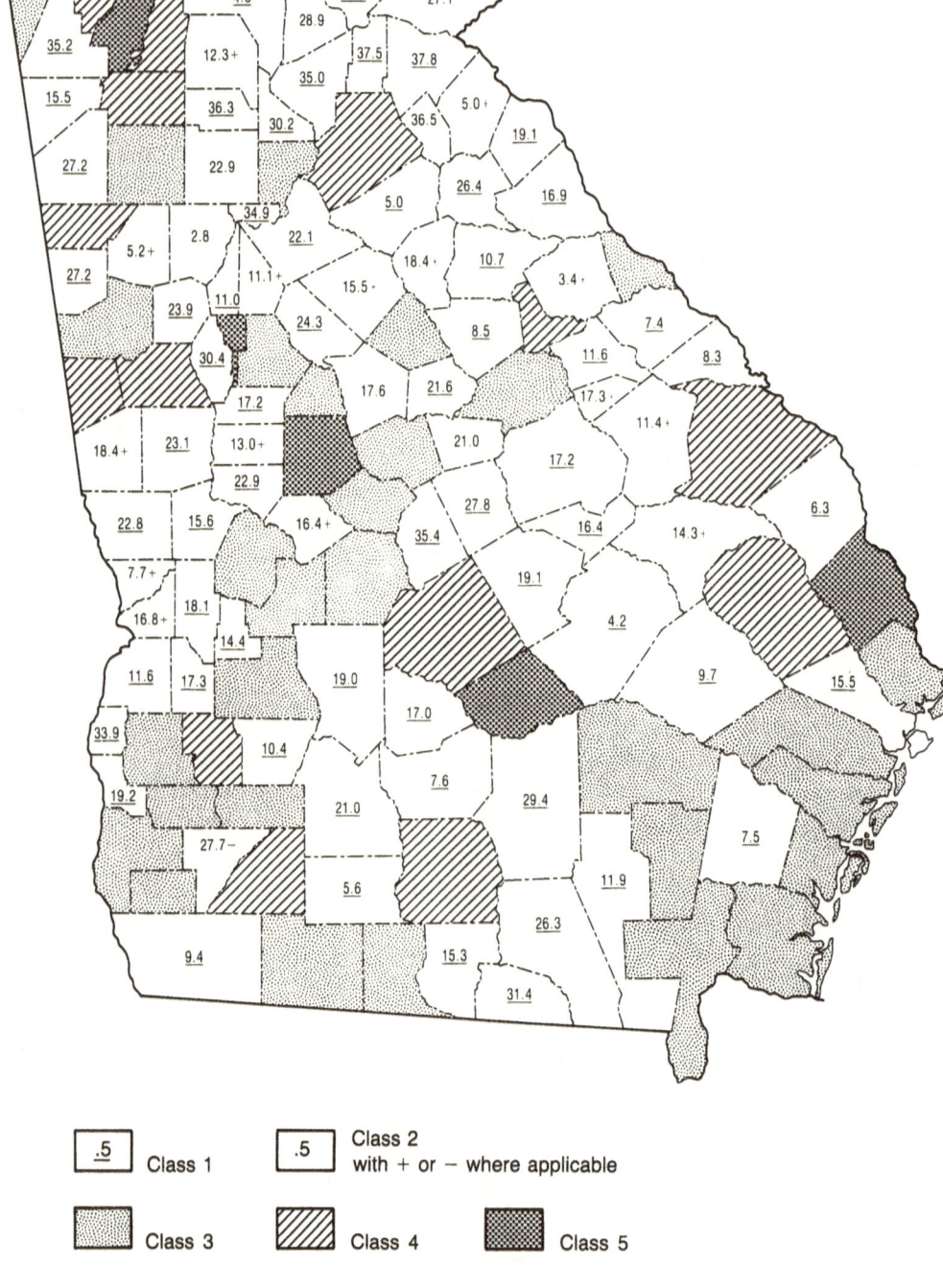

ysis. The patterns found in the two mountain regions seem reasonable and roughly conform to what our hypotheses would lead us to expect. Within the Ridge and Valley, the values tend to be lower in Floyd and Chattooga, which were relatively rich in agricultural valleys and had moderate concentrations of slave labor, than in the more mountainous Walker, where slave concentrations drop off sharply. The higher value of Walker more closely resembles the lower ridge and hill counties of the Blue Ridge region, where the highest values in the state are achieved on the far side of the ridge. The extraordinarily low values of Fannin and Gilmer underscore the importance of interpreting short-method values alongside estimates of types 3 and 5 and of unusually high reportings of "farm laborers." Rough estimations of these supplementary indicators should be compiled simultaneously with the short-method data. In this case it will be seen that Fannin reported a large number of farm laborers, while the returns for Gilmer included a large proportion of type 5 entries and a moderate number of type 3.[4] Downward outliers may be recalculated with ease when farm laborers pose the only problem; types 3 and 5, being more difficult to detect and count, pose a more formidable obstacle for short-method procedures and are probably not worth the effort to correct, at least when the regional pattern is sufficiently clear.

While the short-method estimates produce suitable downward flows as one moves southward through the hill country, they fail completely to delineate the upper Piedmont as a region with tenancy characteristics different from those of the more staple oriented counties of the lower middle to western Piedmont. The long method estimates had similarly failed to detect such a distinction except at level IV. Since the distinction seems to be a real one, and is one that conforms well with what is known about production patterns within central and upper Georgia, we take these values to be pointed reminders that the short method has its limitations. Having said this, we go on to point out that the method produced suitably lower tenancy rates on the older lands of the eastern Piedmont than on the fresher lands to the west. The values in counties falling principally in the Sand Hills, such as Glascock and Crawford, are too low. Our examination of such counties at the militia-district level indicates that substantially higher rates of tenancy may be expected on the poorer lands of the Sand Hills than on the more productive cotton lands to the north and south. Appendix B shows that both Glascock and Crawford counties had significant numbers of type 3 records, however, warning the investigator that the rates are downwardly biased. The Central Cotton Belt displays suitably low rates on the older lands of the east, matching the pattern above the fall line. Values increase as one moves westward to Twiggs, matching the pattern noted in the long method. The values in the southwestern sector of the region increase as one moves southward, but also fail to attain a sufficiently high level in Decatur. The pattern of relations between southwestern Georgia, the Limesink, and the southwestern Pine Barrens are approximately what was achieved through the long method. Within the Pine Barrens the distinction between the new and old lands found by the long method does not entirely hold and may be unfounded. Emanuel may be biased upward somewhat due to the influence of Limesink along its northern boundaries. Even more interesting is the pattern that emerges on the Palmetto Flats along the western edge of the Okefenokee Swamp, where markedly upward values occur in Coffee, Clinch, and Echols counties, with even the swamp-dominated county of Ware displaying an unexpectedly high tenancy rate. It is evident that this subregion within the Pine Barrens deserves special treatment.

It seems reasonable to conclude, then, that the short method generally produces estimated tenancy rates having values that fall within the minimum and maximum range of the long method. Those counties yielding values falling outside that range are frequently identifiable as outliers within their region, or, as in the case of Decatur, they produce values that are flagged as improbable by the anomalous relations they create between regions. Short-method values must be interpreted in conjunction with estimates of types 3 and 5 records and farm laborers. Types b and c records must also be deleted from computations to further ensure comparability between counties. Most gross differences between regions and subregions will then emerge, though some will be missed; however, a basis will have been provided for an informed selection of sample counties that may usefully be subjected to a more intensive analysis using the long method.

If the validity of the long and short methods in establishing estimations of tenancy rates is accepted, and if the regional distributions of tenancy described above seem reasonable, it will be useful to compare the results derived by these methods with the tenancy rates offered for Georgia by Frank Owsley in 1949. Table 6.3 gives Owsley's results alongside our own and provides the breakdowns needed to establish his procedures and to test their reliability. It is clear from examining the first two columns that Owsley in every case included in his denominator a substantial number of farm operators in addition to those enumerated on schedule IV. Since three of his counties were included in our sixteen-county sample, we are able in column 3 to add the precise number of types 4 and 5 entries found on schedule I. While it is quite apparent that he simply miscounted in Harris, the close approximation of totals in Greene and Houston lead clearly to the conclusion that Owsley included both types 4 and 5 in his total farm operator counts.[5] Our totals for Houston County also make it evident that Owsley included types b and c in his totals and made no effort to standardize his data. Column 5 presents tenancy estimates that incorporate column 3 totals in the denominator and our counts of types 2, 3, and 5 in the numerator. The close coincidence in tenancy values again suggests that Owsley followed precisely this procedure. His procedure therefore coincides with our level III procedure, with the important exception that he fails to standardize his data by excluding types b and c. Columns 6 through 8 present our estimates at levels I, III, and IV. It will be seen again that the Owsley estimates may be expected to conform closely to our level III estimates, except where he has miscounted or where the number of types b and c records is high. His miscount in Harris tended to offset his inclusion of a large number of types b and c, resulting in a fairly good percentage;[6] while the inclusion of so large a number of types b and c in Houston, falling as they so often disproportionately do among the tenants, drove the percentage unwarrantedly upward. The tenancy rate for Houston, as derived by Owsley, thus fell marginally outside the established range. In the case of Houston the deviation was not a serious one, but a brief reading of appendix A will reveal that contamination by types b and c can be far more significant in some counties and peculiarly affects tenants. The extent of such contamination should be determined before any reliance is placed on the Owsley rates.

Column 9 compares the Owsley rates with our results derived by the short method. Since the rates compiled by Owsley for Georgia and Alabama are approximately comparable to our level III estimates, because comparable procedures were followed, the Owsley estimates can provide another check on the reliability of the short method in those Georgia counties included by Owsley but excluded from our

Table 6.3

COMPARISON OF TABULATIONS AND RATES (%) DERIVED BY OWSLEY AND BY OUR LONG METHOD AND SHORT METHOD, 1860

County	(1)	(2)	(3)	(4)	(5)	(6)	(7)	(8)	(9)[c]
Elbert	—[a]	573 (5)		—[a]					16.9
Floyd	1,206	1,097 (58)		36.48					27.2
Forsyth	1,074	Class 3		33.05					Class 3
Franklin	941	628 (13)		45.59					5.0+
Glynn	109	Class 3		15.60					Class 3
Gordon	1,045	Class 4		40.57					Class 4
Greene	474	412 (2)	480 (43)	16.46	16.67	7.6	16.7	17.2	8.5
Hall	1,121	Class 4		33.10					Class 4
Harris	934	894 (51)	993 (75)	25.16	28.50	19.1	26.3	27.6	22.8
Heard	674	Class 4		30.27					Class 4
Henry	—[a]	Class 3		—[a]					Class 3

County	(1)	(2)	(3)	(4)	(5)	(6)	(7)	(8)	(9)	Class 3
Houston	721	652 (83)	720 (59)	21.08	21.39	10.6	18.7	19.6		15.3
Lowndes	405	359 (0)		22.72						4.2
Montgomery	340	307 (0)		7.06						9.7
Tattnall	551	467 (5)		21.23[b]						
Wayne	241	187 (1)		24.07[b]						7.5

Source: Frank Lawrence Owsley, *Plain Folk of the Old South* (Chicago, 1965 [first published, 1949]), 155, 171, 177.

Explanation of columns:
(1) "Heads of Families engaged in Agriculture," Owsley's count.
(2) Our count of total schedule IV entries. Figures in parentheses are the number classed as type b or c.
(3) Our count of total schedule IV entries (including types b and c) plus types 4 and 5 found on schedule I. Figures in parentheses are the number that are type 5. Only data for the sixteen county sample are available.
(4) Percentage landless according to Owsley.
(5) Our calculation of types 2, 3, and 5 as a percentage of total schedule IV entries (including types b and c) plus types 4 and 5.
(6) Our estimate at level I by the long method.
(7) Our estimate at level III by the long method.
(8) Our estimate at level IV by the long method.
(9) Our estimate by the short method.

[a] Owsley was unable to locate the 1860 census for these counties.
[b] Derived from tax lists, Owsley being unable to locate the census returns. The tax list for Wayne is 1862, during the war.
[c] Class 1 rates are underscored. For a definition of short-method classes, see chapter 2.

sixteen-county sample. Six counties can be additionally compared in this fashion. The short-method estimate for Floyd provides a good match with Owsley's data, if one notes that over 5 percent of the schedule IV entries are types b and c, thus inflating Owsley's result. The very low estimate for Franklin, which we have placed in class 2, is explained by reference to appendix B, where a high number of type 5 and a moderate number of type 3 records are noted. Lowndes is simply low, but short-method estimates are expected to produce rates that are low but above the minimum. The more serious discrepancies are encountered in Tattnall and Wayne, two Pine Barrens counties, the latter lying partially on the coastal Savannas. These are the only two counties in Georgia that Owsley computed from county tax lists rather than from census schedules. It is surprising that he did not question more seriously the validity and comparability of his results. Montgomery and Tattnall are contiguous counties, both lying along the north side of the Altamaha River and both having similar soils and agrarian regimes. It is not plausible that they should have such radically different tenancy rates. The same conclusion must apply, with marginally less force, to Wayne, which lies to the southeast. The results of Owsley and his students in those counties where tax records are employed must remain suspect until more is known about the reliability and comparability of their documentation.[7]

Unfortunately the procedures followed by Herbert Weaver in Mississippi differed significantly from those of Owsley, his mentor, and of Blanche Henry Clark in Tennessee and do not produce tenancy rates that are at all comparable to our level III estimates. Appendix D presents a regional breakdown of the results of Owsley for Alabama and Georgia, of Harry L. Coles, Jr., for Louisiana, of Weaver for Mississippi, and of Clark for Tennessee. A description of the different procedures employed by each may be found in chapter 2. Our recalculated Mississippi proprietor estimates given in parentheses undoubtedly mirror tenancy rates higher than our level III estimates because, although they almost certainly include records that are not in fact tenants, they also exclude type 4 records from the denominator. Our recalculated estimates probably fall between levels III and IV. The result of these recalculations is to bring Mississippi into line with regional differences within Georgia (except for the questionable result in the Georgia Pine Belt). Tennessee would seem to be a state peculiarly marked by high rates of tenancy. Very little can be said of Louisiana, because Coles's procedures remain unclear to us.

Bearing in mind the limitations of Owsley's and his students' procedures, the data gathered lend considerable support to the thesis that antebellum tenancy was concentrated principally outside the cotton lands, reaching its highest levels in the upland regions. But tenancy was far from negligible in the cotton regions, even in those where production was most intensive. The rate in 1860 fell below 19 percent only in the Alabama Black Belt. It is unlikely that the rates computed by Owsley himself are too low, at least where census records were employed. On the contrary, it is possible that his rates and those of his students to some extent underestimate regional differences by failing to achieve level IV estimates in any sample counties where a close examination might suggest level IV as more appropriate than level III. As we have seen, level IV estimates are especially likely to enlarge tenancy rates in counties outside the cotton belt. Nor is it clear that the counties selected for the Owsley samples yield reliable averages for these vast and sometimes significantly heterogeneous regions. But despite their generality, the Owsley-school statistics do support the plausibility of the patterns of tenancy we have found in Georgia, and it is one of the more curious mysteries of modern historiography that these patterns have been so widely neglected. One important exception to our patterns must be noted, however. No Pine Barrens region outside Georgia fails to have less than 27 percent tenancy in 1860, a figure approaching our results in the Palmetto Flats. If the results for these regions are reliable at level III, then our own investigation of tenancy in the Georgia Pine Barrens is inadequate and has seriously underestimated the actual rates. The rate for Emanuel, achieved by the long method, is at best only barely plausible within such a context of higher actual rates, but Emanuel is the only county in the older eastern sector of the Pine Barrens we have tested by the long method. It is possible that short-method procedures have failed in this region and have produced differences between the older and newer Wiregrass and the Palmetto Flats that are not so large as we have portrayed. A more complete investigation may show the county tax lists to be more reliable than we are presently inclined to concede.[8]

Many counties analyzed by the long method were reported within militia-district subdivisions, and we hoped to be able to provide a detailed analysis of this micro level in order to bring our interpretation of tenancy distributions into a tighter relationship with the variables. While such a discussion has largely been frustrated, principally because of flaws we have discovered in the modes of reporting, it is nevertheless apparent from appendix C, as well as from appendix H, that considerable variations in structural characteristics may be found among militia districts of a single county. An examination of Baldwin and Twiggs counties will quickly reveal that sharply higher tenancy rates are found in Sand Hills districts. In Baldwin County all of district 322 and most of districts 321 and 115 lie in the Sand Hills. District 355 in Twiggs is entirely Sand Hills, and district 356, while presenting a more complex problem, may be considered Sand Hills for our purposes.[9] Similarly, in Harris County, the highest tenancy rate, that in district 1186, is found on the worst lands of the county. The lowest tenancy rates, on the other hand, are always found on good lands, but normally not on the best. The best lands in Twiggs County lay in district 396, which produced a high tenancy rate, unusual in a banner district. In Harris the best lands were at the headwaters of Mulberry Creek, in district 934 (the banner district), and in the Pine Mountain Valley districts 786 and (in part) 703. The lowest tenancy rates in Harris occurred on more intermediate lands stretching diagonally across the entire length of the county in a northwesterly to southeasterly direction.

We undertook a preliminary data analysis of tenancy and a few other variables at the militia-district level in these three cotton counties—Baldwin, Harris, and Twiggs—comparing rank orders and bivariate correlations. Some districts in these counties produced outliers, and it is clear that such districts require special treatment as smoothing occurs in more sophisticated analysis. For example, district 320 in Baldwin is a small district and contains Milledgeville, the state capital. Such urban districts do not invariably create outliers; sometimes they drive the linear regressions toward higher coefficients. But they do so for the wrong reasons, and it is wiser to withdraw their urban contamination from the general rural analysis and to reserve it for special examination. District 321, in Baldwin, also created outliers in some variables because of the presence of a significant number of unusually wealthy planters who resided on small home farms in Scottsboro, a popular local summer residence known for its sulphur springs. The banner districts in Harris and Twiggs also proved to be outliers with respect to tenancy. Bivariate correlation coefficients leapt upward when these districts were withdrawn. But while preliminary data analysis proved in many ways unsatisfactory, it seems reasonably safe to conclude that the highest tenancy rates within cotton counties tended to occur on the worst lands. Apart from these worst districts, we found some tendency for tenancy to decrease as the quality of soil and such variables as mean improved acreage and mean personal wealth increased, but the relationships were weak. They were strong in Baldwin when the two extreme outliers were removed, but the number of remaining districts was small. Significance levels dropped sharply as we moved from a less cotton intense county, Baldwin, through Harris to a more cotton intense county such as Twiggs.[10]

It is clear that sharp differences in tenancy rates and in all other variables derived from the census occur between militia districts. It is also clear that the range of variation is striking. A well distributed range of 26.6 among districts in Harris County is not to be sniffed at. Even outside the cotton belt, where the relative uniformity of land quality and the absence of staple production would lead one to expect a more uniform distribution of tenancy, we find a range among districts in Coffee extending to 18.8. With ranges such as these occurring between districts, it would appear to be desirable to work at militia-district levels wherever possible in order to develop more sensitive indicators for multivariate analysis. County-level statistics unavoidably cloud a richness of variation. Unfortunately, variations in the mode of reporting, along with procedures imposed by the census forms themselves and the instructions that accompanied them, often make militia-district statistics either unreliable or misleading. They do not measure what they purport to measure.

The deficiencies we have detected in data obtained at the militia-district level can be traced to two principal causes. First, a comparison of total farm acres reported in the census for each militia district and the actual acres contained within that district reveals that some districts were severely overreported while others were severely underreported (several of these comparisons are shown in table 6.4). We have been unable to find reliable modern reportings of militia-district acreages; however, measurement of those districts sufficiently geometric in shape on a modern one-inch-scale highway map provide approximations that are sufficient for our purposes, especially when the discrepancies are great.[11] It is clear that the total acres reported by the census for all districts closely approximate the actual surface area of each county. Acres are not being under- or overreported in the counties as a whole. But it is equally clear that—unless militia-district boundaries have been grossly altered over the years, which in

Table 6.4

COMPARISON OF PRESENT SURFACE AREA AND REPORTED FARM ACREAGE IN SELECTED MILITIA DISTRICTS, 1860

County	Militia District	Estimated Present Surface Area (Acres)	Reported Farm Acreage, 1860	Farm Acres as Percentage of Estimated Surface Area	Present Surface Area (Acres)	Total Reported Farm Acres, 1860	Total Farm as Percentage of Total Surface
Baldwin					169,600	159,826	94.2
	319	20,992	25,245	120.3			
	322	21,376	13,272	62.1			
Harris					302,720	296,089	97.8
	695	22,144	26,210	118.4			
	707	26,240	19,087	72.7			
	781	20,864	31,362	150.3			
	782	15,104	17,286	114.4			
	934	17,664	21,452	121.4			
Twiggs					233,600	232,409	99.5
	326	17,984	17,487	97.2			
	354	32,000	19,413	60.7			

these cases is improbable[12]—the distribution of reported census acres between districts bears no relation to the actual surface area of many districts. The explanation for such discrepancies seems to lie in the failure of the census to report the acres of a proprietor in those districts where he held property but did not reside, and to report all the acres of the proprietor in the district of his residence irrespective of where they lay. Such a procedure was required by the census instructions. The census forms also imposed such a procedure on the enumerator unless he was prepared to enter a proprietor on schedule IV in each district where he held property. We found no such entries in our work. It is inevitable, then, that some number of proprietors reported acres in one district that actually lay in one or more other districts; it is probable that the incidence of scattered holdings rose in direct proportion to the size of a man's holdings and the labor at his disposal. It is little wonder under these circumstances that simple linear relations based on means and percentages largely failed to materialize in analysis at the militia-district level or that the failure was most acute in those counties where the largest and most wealthy planters resided.

Data obtained at the militia-district level may well reflect real differences between districts, and for some purposes they may be more useful than data at the county level, which, as we have seen, have their own plentiful deficiencies. But bear in mind that militia-district statistics may be sufficiently correct in their relative magnitudes to be useful in rank orderings, for example, but incorrect in their precise magnitudes. In some cases even rank orderings may be distorted. Linear analysis poses even higher risks. Even more important, the differences between districts that preliminary statistics seem to reflect may not lie in the differences they purport to measure. For example, a district with very poor soils may report very high tenancy rates and low farm size and personal wealth, even among type 1a proprietors. One would conclude that there was only a smallholder presence in such a district. Such a conclusion may be unwarranted, however, if it can be determined that the acreage in that district is substantially underreported. If such underreporting is found to occur, then it is quite possible that the missing acres are under the ownership and control of one or more largeholders, and that even a sizable portion of the tenants of that district hold their farms from such persons. It is clear that a political study attempting to relate voting patterns with landholding could go seriously astray on such points, and so too could a study of landholding itself or of the relations between landholding and capital. Tenancy rates as such are less vulnerable to absentee proprietorship than most statistics, if (and only if) one can assume that the number of absentee proprietors is small in relation to the tenants.

The second cause of deficiencies that we have detected in data at the militia-district level is similar to the first but pertains to reportings of personal wealth. Like acreages on schedule IV, personal wealth is assigned on schedule I within the district of residence. But unlike acreages, personal wealth is not even confined to the county but may be physically located in any number of counties within the state and even beyond. By far the most significant element in personal wealth throughout the southern states before the war was slaves. Because slaves tend to go with land, they tend to scatter as acres scatter; but the problem is even more complex. Not all slaves were directly used on the lands of the owner; many were hired out. And many were sent beyond the bounds of the county in which the owner's personal wealth was reported. It is also probably true that they were disproportionately concentrated on the best lands. A careful use of slave schedules should therefore accompany this variable if it is relied on, not so much to estimate the magnitude of a person's wealth as to locate it spatially. It is not yet clear, however, that all slave schedules properly assign slaves to militia districts, even if they do properly assign them to counties. Our present impression is that many slave schedules suffer from the same militia-district deficiencies as schedule IV.

Our conclusion is that such statistics as mean improved acres, percent improved, value per acre, and bales per improved acre are more reliable at the county level than at the militia-district level for the reasons just discussed. It is also probable that crop and livestock values should be treated at the county level for the same reasons. Even at the county level, reportings of improved acres, and all associated statistics, may be misleading within counties, and lack comparability between counties, because the Census Office failed to provide a clear definition of what was to be understood by the term *improved*. There is some reason to believe that in some counties exhausted and abandoned lands lying out were included in the improved category, while in others they were excluded. Such inconsistency would have a differential impact on counties of old and new settlement and on any attempt to compare values between census years. Personal wealth as recorded on schedule I clearly introduces a more substantial bias at the militia-district level than at the county level. The bias is all the more evident when one remembers that while "personal" includes wealth outside the county, only largeholders are likely to have sufficient outside wealth to seriously disturb the county profile. But even moderate holders may have acres and slaves beyond the boundaries of their militia district, and the presence of even a single very large holder can easily swamp the mean and distort the district. The advantage of the militia district, which is its small size, is also its disadvantage. On the other hand, given all the above reservations, differences that are real may be expected to emerge from analysis at the militia-district level, even though the precise nature and magnitude of those differences may be concealed, at least so long as research is restricted to census records. A use of court and land records, along with plantation correspondence and rentals, could clarify most or all these problems. Meanwhile the data given for tenants themselves—acreages and farm value where available, along with values for livestock and crops—should be as reliable at the militia-district level as at any other, since tenants were overwhelmingly smallholders and unlikely to have additional holdings beyond the boundaries of their district.

Despite the monumental deficiencies in the data, tentative conclusions regarding the distribution of tenancy within Georgia cotton counties in 1860 seem warranted. The tentative nature of our conclusions must be sharply emphasized, however. We begin with the assumption that tenancy is the best alternative to slave labor and, being a less expensive form of labor, is characteristically employed either when soils or conditions do not justify slave labor or when slave labor is lacking. A supply of appropriate tenants sets up further conditions for the supply and demand curves, but the curves are complicated by the term *appropriate*, which masks a variety of additional variables. There must be a suitable mix of labor, land, and other forms of capital such as livestock, seed, and guano. An analysis on the basis of census variables alone, assuming such a data base to be a refined and reliable version, is therefore a complex undertaking at best. Without undertaking so complex an analysis, it seems clear that the highest tenancy rates within cotton counties occur on the poorest lands, where slave investment makes a poor return. Such a conclusion conforms well with our intercounty analysis, where tenancy rates also rise as one proceeds to the margins of the staple economy and beyond, with the qualification that the new lands–old lands distinction cuts across such a generalization. Our correlation coefficients for Baldwin and Harris counties suggest an inverse relation between tenancy and land quality on lands of intermediate value, but the results are too inconclusive to rely on—and the data are too distorted, as we have shown, to believe the highest coefficients even were they obtainable.[13] Tenants with personal capital were to some extent able to compete for holdings on prime lands, but such lands were principally reserved for slave labor, since the returns on expensive labor investment were high. Downward pressure was therefore exerted on tenancy in banner districts owing to high returns on labor and the relatively few tenancy candidates with sufficient capital to work a prime plantation to capacity. Once again, such a conclusion fits well with our intercounty analysis, where low tenancy rates are found in such banner counties as Houston.

If we shift our perspective back to the county and state levels, we may now also draw some conclusions about the farm size and personal wealth characteristics of antebellum tenants, or at least of those who possessed their land and legally owned their crops and were therefore most likely to be reported on schedule IV. When viewed from this larger perspective it is quite clear that in the cotton belt the size of tenants' holdings, as well as their personal capital, increased inversely with the value of the land. Some tenants, especially in banner districts, were substantial farmers and may have been operating holdings organized as small plantations. A reference to table 6.5 will show that at least one tenant in Baldwin and one tenant in Wilkes operated holdings in excess of 1,000 improved acres. But in the counties presented in this table—which represent the old counties of the lower eastern Piedmont, the intermediate counties of the lower central Piedmont, and the newer counties under heavy clearance in the southwestern Central Cotton Belt—the preponderance of types 2a and 3a farms were between twenty and forty-nine improved acres, except in Wilkes, where the type 3a farms raised the median to the fifty to ninety-nine-acre interval. Outside the cotton belt the size of tenant farms was normally well below fifty acres; and in a county like Towns, where acres for tenants are not given, production figures for many holders cause one to wonder how subsistence was maintained. Tenants on the margins and well outside the staple economy were doubtless engaged in more than farming, supplementing their diets and their incomes by

Table 6.5
TENANT FARM SIZE DISTRIBUTION, 1860

County	Type	Farms	Improved Acres							Totals
			3–9	10–19	20–49	50–99	100–499	500–999	1,000+	
Baker	2a	Number	1	1	21	7	2			32
		Percentage	3.1	3.1	65.6	21.9	6.2			100
	3a	Number			4	3				7
		Percentage			57.1	42.8				100
	All[a]	Number	2	1	52	37	65	40	10	207
		Percentage	1.0	0.5	25.1	17.9	31.4	19.3	4.8	100
Baldwin	2a	Number	13	12	20	9	2			56
		Percentage	23.2	21.4	35.7	16.1	3.6			100
	3a	Number	4	7	9	1	2			23
		Percentage	17.4	30.4	39.1	4.3	8.7			100
	All[a]	Number	30	33	78	66	121	17	1	346
		Percentage	8.7	9.5	22.5	19.1	35.0	4.9	0.3	100
Wilkes	2a	Number		2	3	2	4			11
		Percentage		18.2	27.3	18.2	36.4			100
	3a	Number		1	5	10	9	1	1	27
		Percentage		3.7	18.5	37.0	33.3	3.7	3.7	100
	All[a]	Number	1	5	30	57	202	69	29	393
		Percentage	0.2	1.3	7.6	14.5	51.4	17.6	7.4	100

[a]Totals per interval for all landholders are taken from the published aggregate census. Our own count of farms reporting 3 or more improved acres comes to 2 less in Baker County and 14 less in Wilkes. In the case of Wilkes, differences could arise because of our aggregation of manager entries under proprietor. Our count in Baldwin is also lower, but one page of approximately forty records was missing from the microfilm but seems originally to have reached the Census Office.

hunting and fishing and, in the Pine Barrens, by lumbering and turpentine production.

A fuller data series can be obtained for the personal wealth of tenants, since it is purportedly uniformly reported on schedule I. Appendix F shows that the majority of tenants were men of little or no means. Easily the wealthiest group of tenants were to be found in Houston, a prime cotton county with a low tenancy rate. The poorest group of tenants within the cotton counties lay in Gwinnett on the outer margin of the staple region in the upper Piedmont. The old but prestigous counties of the eastern lower Piedmont not only produced their share of wealthy tenants but, in the case of Wilkes, produced them in higher than normal proportions. The personal wealth data, like those for farm size, suggest a wealthier class of tenant in this subregion. But with the exception of a few banner counties and districts, the pattern suggests that the overwhelming majority of Georgia tenants in 1860 were men and women operating small holdings with little or no personal capital. They operated principally on the spatial margins of the cotton economy or outside it, yet those who participated in the cotton economy—and the large majority in the cotton counties reported bales—did so at intensity levels only slightly below those of neighboring proprietors. Their output was small and relatively insignificant, but their participation was high.

7

The Transition to 1880

THERE IS NO DOUBT THAT southern tenancy increased substantially following the Civil War, and there is nothing in our findings that can alter such a conclusion. Ransom and Sutch have calculated that tenancy in the South as a whole in 1880 constituted almost 39 percent of the landholding population. In the Old South alone—including Virginia, Georgia, and the Carolinas—the rate was almost 38 percent. Within the Old South division, tenancy rates in the Atlantic Seaboard counties were 38 percent, on the Atlantic Coastal Plain (which includes the Wiregrass and Central Cotton Belt counties) they were at 41 percent, and on the Piedmont Plateau the rates reached their highest regional value of 42 percent. Only in the Appalachian Hills area did the rate fall as low as 25 percent.[1] While some of these 1880 rates can be matched in 1860, they can be matched only in regions that were both spatially and economically on the margins of antebellum Georgia. They cannot be matched in regions that were significant for staple production. And the 1860 figures for Georgia cannot collectively achieve the levels Ransom and Sutch have calculated for the South as a whole.

Nevertheless it is also clear that the rates of tenancy in Georgia on the eve of the Civil War were sufficiently high to preclude any interpretation portraying whites as having "fallen into tenancy" under Reconstruction.[2] The rates of tenancy increased in some areas, declined in others, and increased substantially overall; but the magnitude of the overall increase, and the complexity of its spatial patterns, is quite different from that previously portrayed by historians. It is similarly unnecessary to portray postbellum tenancy—perhaps even sharecropping—as a postbellum invention without precedent, in the South at least, devised hurriedly and mysteriously as a structural alternative to slavery.[3] The statistics, however provisional they may be, seem clearly to indicate that while tenancy increased after the war, the scale of change was not so dramatic or so distributed as historians have hitherto been forced to believe. What is dramatic, and what did change, is that a formerly white monopoly over a valuable and scarce resource—land—was opened to black competition.[4] We believe we have thus been able to identify an additional economic basis for postbellum white racism.

A discussion of postbellum racism is well beyond the bounds of our present discussion, however. In this chapter we shall attempt to delineate the larger contours of the transformation that occurred in tenancy rates between 1860 and 1880, and we shall interpret these rate changes by relating them to changes in the distribution of white and black population, to patterns of increase in improved acreage, and to regional shifts in levels and efficiency of cotton output. The discussion will proceed on the basis of the preliminary estimates we have derived for 1860, coupled with data easily derived from the aggregate census for 1880. Neither set of data is fully sufficient for such a discussion; but the data may be considered sufficiently reliable if the conclusions are treated as preliminary and tentative. We shall begin by directly comparing regional distributions of tenancy rates derived for 1860 by the long and short methods (see maps 6.5 and 6.6) with those computed from the aggregate census of 1880.

County tenancy rates for 1880 are presented on map 7.1. The briefest study of this map reveals sharp regional differences. A comparison of this map with those presenting our 1860 estimates will similarly show the contours of regional alteration that occurred in tenancy distributions in a society undergoing major spatial shifts in population and staple production during the transition from an economy based on slavery and tenancy to one based solely on tenancy. Tenancy rates increased in the upper central Piedmont by about 30 percent, rising from the low thirties to the low forties. In the lower western Ridge and Valley and in the lower central and western Piedmont, as well as along the coast,[5] the 1880 rates were roughly double those of 1860. In the southwestern counties of the Central Cotton Belt the rates rose by 100 percent to 150 percent, and the counties around Houston rose more steeply by 200 percent to 300 percent. The sharpest rise in tenancy

Map 7.1. Tenancy (Cash Renters and Share Renters) as Percentage of Farm Operators, 1880

Notes: The state mean is 44.8 percent. See 1880 map on page xxi.
Source: U.S. Census Office, Tenth Census [1880]. *Report on the Production of Agriculture in the United States at the Tenth Census* (Washington, D.C.: GPO, 1883).

rates in the state occurred on the older lands of the eastern Piedmont, where increases ran as high as 400 percent. Little change occurred in tenancy rates in a few counties of the upper western Piedmont around Haralson or in the extreme southwest (Thomas) or in the western Limesink (Worth). In several subregions the rates declined. The decline was sharp, as much as 50 percent, in the far ridge counties of the Blue Ridge region and on the newer lands and Palmetto Flats of the Pine Barrens. The near ridge counties, such as Habersham, and the older counties of the Pine Barrens region experienced a modest decline of about 5 percent.

While it is apparent that important structural alterations in tenancy were occurring between 1860 and 1880, it is not apparent from tenancy rates alone what the precise nature of that change might have been. An examination of table 7.1 will help to clarify what has happened. The table compares the actual numbers of proprietors and tenants in our sixteen sample counties in 1860 with the actual number of proprietors and tenants in 1880, distinguishing renters and farmers on shares. The dramatic increases in tenancy rates experienced on the old lands of the eastern Piedmont were produced by extraordinarily high increases in absolute numbers of tenants and by moderate increases in proprietors. Although it is true that the magnitude of increase in the rate was principally a function of the low rate of tenancy prevailing in this subregion before the war, it nevertheless remains that this old and aristocratic cluster of counties experienced a more dramatic structural change in its landholding following the war than did any other subregion in the state. Sharper real increases in tenancy occurred in counties such as Baker, which experienced a high increase in the absolute number of its tenants but a very low increase in proprietors. All the cotton counties in the sixteen-county sample located in the lower western Piedmont or in the central and southwestern sections of the Central Cotton Belt—Baker, Clay, Harris, Houston, and Twiggs—display a similar pattern to varying degrees. But although the magnitude of their rate increases was high and the structural change considerable, the fact that all these counties had experienced significant rates of tenancy before the war must have lessened the dramatic impact of the change when compared with counties like Greene and Wilkes.

The table also makes it evident that large variations in tenancy rate rise are not due only, or at times even most importantly, to increases in

Table 7.1

COMPARISON OF PROPRIETOR AND TENANT INCREASES, SIXTEEN SAMPLE COUNTIES, 1860 AND 1880

County	Proprietors, 1860 Types 1a + 6a + 4	Tenants, 1860 Level III/IV	Tenants, 1860 Level III/IV (%)	Proprietors, 1880 (with % Increase)	Cash Renters, 1880	Farmers on Shares, 1880	Total Tenants, 1880 (with % Increase)	Tenants, 1880 (%)
Baker	182	91	33.3	207 (13.7)	561	185	746 (719.8)	78.3
Baldwin	246[a]	91[a]	26.2	349 (36.3)	109	608	717 (687.9)	67.3
Clay	252	90	26.3	284 (12.7)	225	90	315 (250.0)	52.6
Coffee	245	103	29.6	476 (94.3)	11	62	73 (−29.1)	13.3
Echols	93	80[b]	46.2	213 (129.0)	5	59	64 (−20.0)	23.1
Emanuel	473	99	17.3	941 (98.9)	39	100	139 (40.4)	12.9
Greene	398	83	17.2	594 (49.2)	425	856	1,281 (1,443.4)	68.3
Gwinnett	843[a]	407[a]	32.6	1,447 (71.6)	58	993	1,051 (158.2)	42.1
Habersham	489	348	41.6	682 (39.5)	13	364	377 (8.3)	35.6
Harris	694	248	26.3	732 (5.5)	513	517	1,030 (315.3)	58.4
Houston	518	126	19.6	551 (6.4)	408	429	837 (564.3)	60.3
Thomas	367	206	36.0	983 (167.8)	170	435	605 (193.7)	38.1
Towns	234	174	42.6	340 (45.3)	13	81	94 (−46.0)	21.6
Twiggs	298	160	34.9	343 (15.1)	204	316	520 (225.0)	60.2
Wilkes	364	60	14.2	554 (52.2)	190	603	793 (1,221.7)	58.9
Worth	230	98	29.9	432 (87.8)	4	185	189 (92.8)	30.4

[a]Downwardly defective because of missing pages on schedule IV. See appendix H.
[b]Fifty-two of eighty were "farm laborers" entered on schedule I only.

absolute numbers of tenants. Differential changes in numbers of proprietors are also important in understanding the rates. Despite the missing forty or so records from our 1860 analysis of Baldwin County, it seems that proprietor increases in the lower central Piedmont may have moderately depressed the rates in this subregion of the old cotton economy. The graded increases in proprietorship as one moves from west to east along the lower Piedmont accords well with suggestions in the literature that land that had been lying out in the older counties during the late antebellum period was now being brought back into profitable staple production.[6] A comparison of proprietor increases with tenant increases becomes even more interesting when one moves to the upper Piedmont. Although substantial portions of the 1860 returns for Gwinnett are missing, and were apparently never submitted to the Census Office,[7] it is safe to assume that the distribution of the change between proprietor and tenant would not be radically different if the records had survived. When compared with changes in counties in the lower Piedmont, structural change in Gwinnett favored tenancy but only to a moderate degree. Percentage increase in proprietorship was almost half that in tenancy. More pointedly, Thomas County, in the extreme southwest, and Worth, in the western Limesink, both experienced large increases in numbers of tenants, but these increases were matched by virtually equivalent increases in numbers of proprietors. The tenancy rates in these counties therefore remained virtually unchanged as well, despite the large increases in tenants.

The comparison between proprietor and tenant increases becomes most instructive in those counties where the tenancy rate declined. In Habersham County, on the southern slopes of the Blue Ridge region, the rate declined from 41.6 percent in 1860 to 35.6 percent in 1880; in Emanuel, an older county of the eastern Pine Barrens, the rate declined from 17.3 percent to 12.9 percent. In each of these counties, although the rate declined, the absolute number of tenants increased, by 8.3 percent in Habersham and 40.4 percent in Emanuel. The decline in the rate was caused by even larger proportionate increases in proprietorship in both counties, 39.5 percent in Habersham and 98.9 percent in Emanuel. It would therefore be incorrect to leap easily to the conclusion that tenants were being drawn off these lands and going elsewhere. Tenancy as such would appear to have been declining, when measured merely as a proportion of landholding. But such a decline must be measured against rates of natural population growth, patterns of geographical mobility, and estimates of intracounty allocations of tenancy as labor before any clear interpretation may emerge regarding the import of this decline.

In three counties of the sixteen-county sample, the absolute numbers of tenants declined. Towns County, lying on the high and relatively inaccessible northern slopes of the Blue Ridge, increased its proprietors by 45.3 percent but lost the same percentage of its antebellum tenants. Its tenancy rate declined from one of the highest in the state in 1860 (42.6 percent) to 21.6 percent in 1880, a figure well below the state mean of 44.8 percent. There is no doubt that counties within the far ridge section of the Blue Ridge, unlike Habersham and other counties on the southern slopes and hills, underwent substantial change in the structure of their landholding following the war. If Coffee is typical of its subregion, a similar change in landholding seems to have occurred in some western portions of the Pine Barrens. But Echols County, lying on the Palmetto Flats to the south of Coffee, suggests caution in generalizing too precisely about this and other subregions. While proprietors in Echols increased by 129 percent and tenants declined by 20 percent, it should be noted that fifty-two of the tenants in 1860 were described as "laborers" and appeared on schedule I only, a mode of reporting that suggests sharecropping. In 1880 fifty-nine of the sixty-four tenants in the county were farmers on shares. It may be that there was a marginal increase in the number of sharecroppers during the twenty years, a sharp decline in renters, and a steep rise in proprietors. While it cannot be doubted that multiple forms of tenancy were present in antebellum Georgia, census records alone will never permit one mode to be distinguished clearly or with any certainty from another. Any investigation that relies on census records alone will therefore be unable to distinguish or even to detect many structural alterations in the allocation of labor, even when the magnitude of those alterations was large.

Maps 7.2 and 7.3 must therefore be only deceptively useful in detecting alterations in modes of tenancy. They purport to reveal the relative distribution of cash renters and renters on shares throughout the state in 1880. Unfortunately there is no reliable comparable data for 1860 that can be derived from the census alone and against which this later data may be compared. One could assume that all records of verified tenants in what we call standard convention (that is, without

Map 7.2. Cash Renters as Percentage of Total Tenants (Statistics), 1880

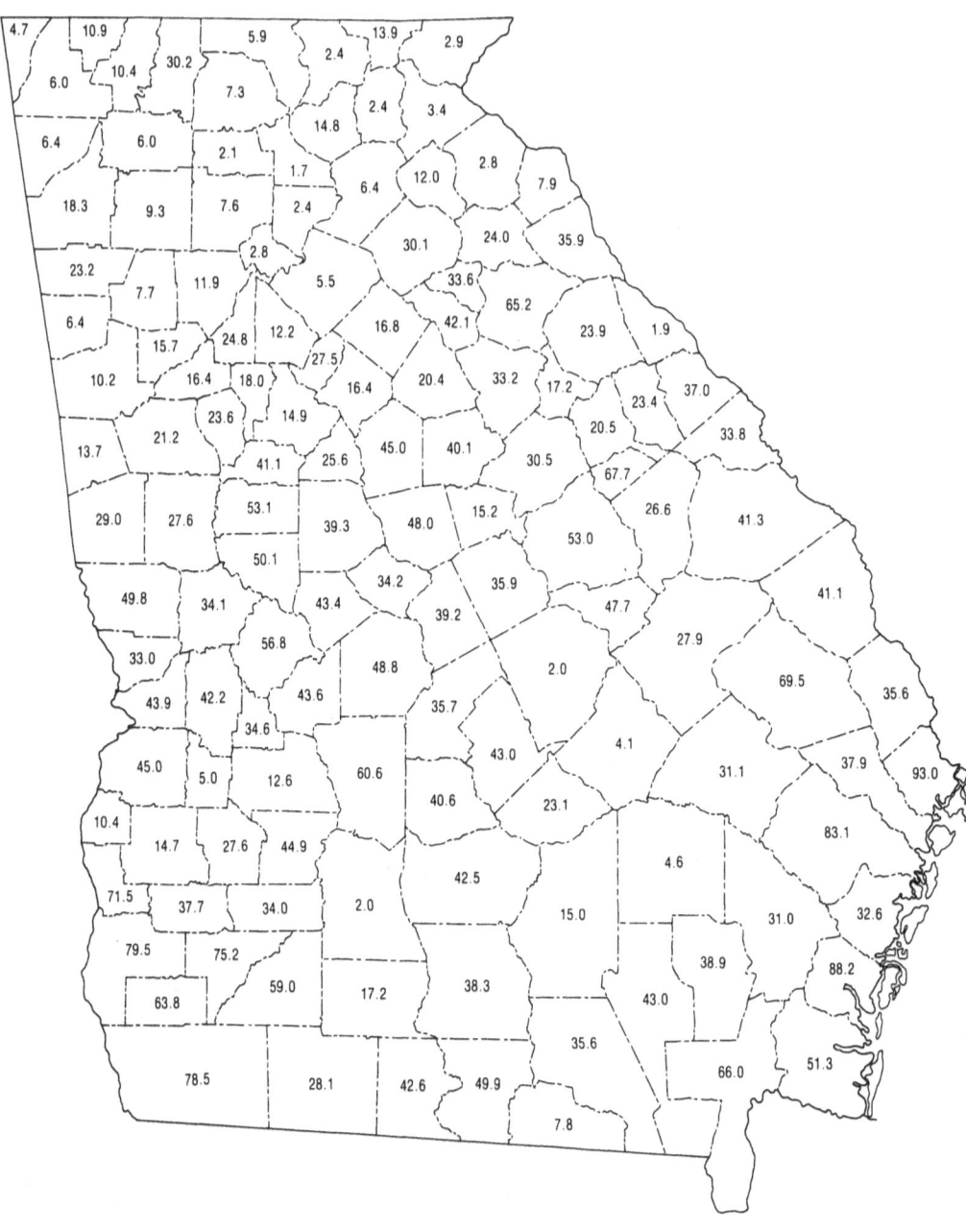

Notes: The state mean is 30.1 percent. See 1880 map on page xxi.
Source: U.S. Census Office, Tenth Census [1880]. Report on the Production of Agriculture in the United States at the Tenth Census (Washington, D.C.: GPO, 1883).

Map 7.3. Cash Renters as Percentage of Total Tenants (Shaded), 1880

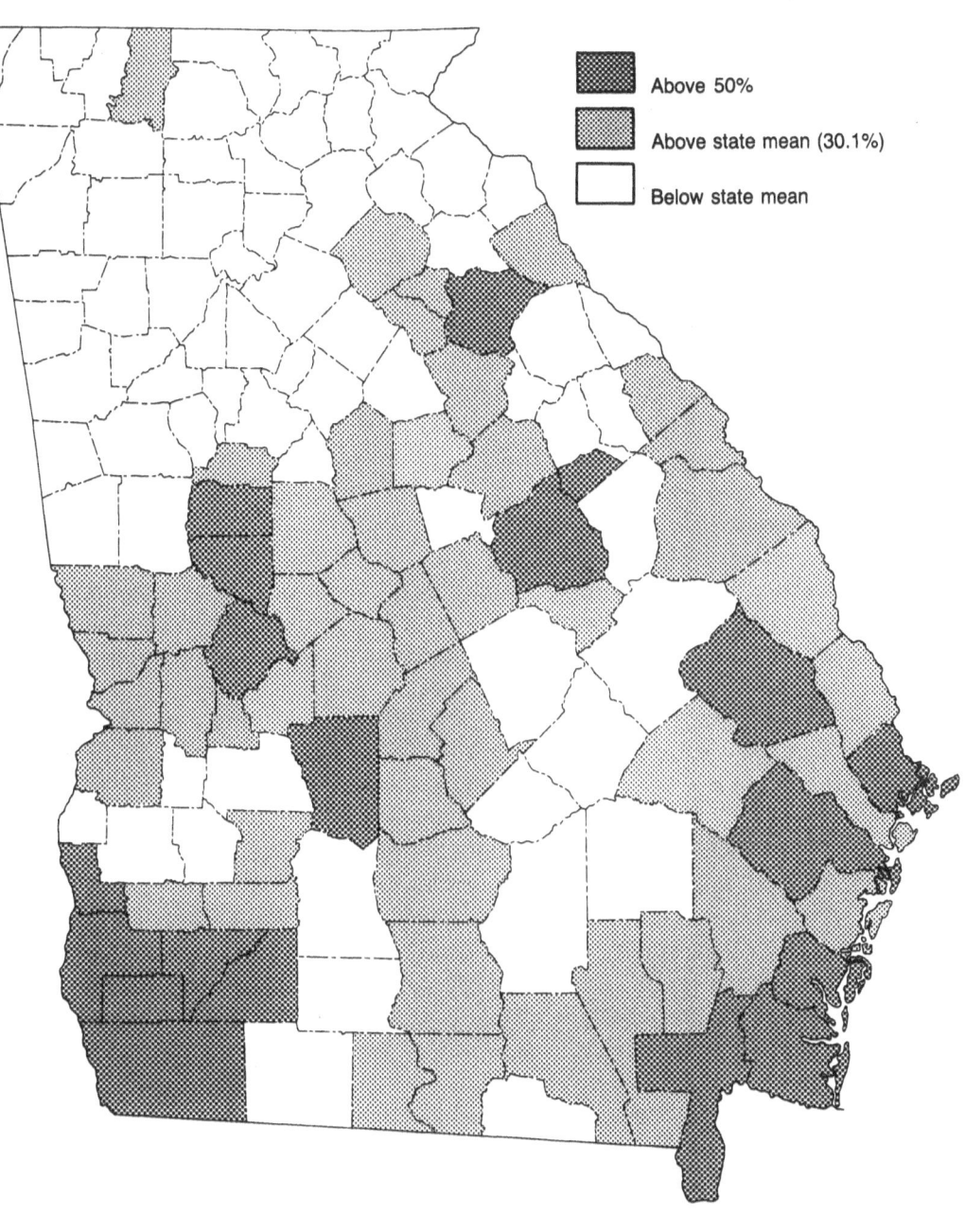

■ Above 50%
▨ Above state mean (30.1%)
□ Below state mean

Note: See 1880 map on page xxi.
Source: U.S. Census Office, Tenth Census [1880]. *Report on the Production of Agriculture in the United States at the Tenth Census* (Washington, D.C.: GPO, 1883).

acreage and farm value but reporting livestock and crops) in 1860 were either cash renters or share renters, since the livestock and crops seem to have been considered their own but the farm the proprietor's; and all forms of tenancy reporting on schedule I only could be considered sharecropping in one of its forms. When full values are reported on schedule IV (our type 3), cash or share renting could be similarly assumed. Such assumptions are probably rash in general, but are extremely rash in any particular application and ought not to be employed in systematic analysis. Brooks has pointed out, on the basis of his own extensive experience as a census field supervisor, that enumerators in late nineteenth- and early twentieth-century Georgia, in accordance with their instructions, commonly grouped share renters on thirds and fourths with sharecroppers, especially in the upper Piedmont and mountains. But renters on thirds and fourths were considered by planters to be in the same category as cash renters.[8] Such confusions may normally be expected to produce enumerator inconsistency. It is therefore far from certain that the neat regional patterns appearing on map 7.3, with the low cash-renter ratios in the mountains and Pine Barrens, reflect true regional differences. If such differences did exist, their actual nature may be masked. County enumerators may have reported with their usual inconsistency; enumerator inconsistency is all the more probable when one remembers that tenancy was not a simple dichotomy. While it may have been a dichotomy in law, it was not a dichotomy that was simple to apply in law or to detect in the field. Even so, the fact that in 1860 Towns reported tenants only in standard convention may lead one to conclude that all tenants in the county were either cash or share renters. This conclusion is difficult to sustain, however, when one notes that Fannin, also a northern slope county within the Blue Ridge, reported its tenants on schedule I only and as farm laborers, and Gilmer in the same region reported its tenants as types 3 and 5 with the latter strongly preponderant. These observations suggest that standard convention, when unaccompanied by other conventions, may indicate that the enumerator is simply ignoring tenancy distinctions.

The import of spatial alterations in the distribution of tenancy may be clarified by comparing these alterations with movements of white and black population within Georgia during the same period. Map 7.4 presents for each county the percentage increase in its white population between 1860 and 1880. Some counties are aggregated be-

Map 7.4. Percentage of Change in White Population (Statistics), 1860–1880

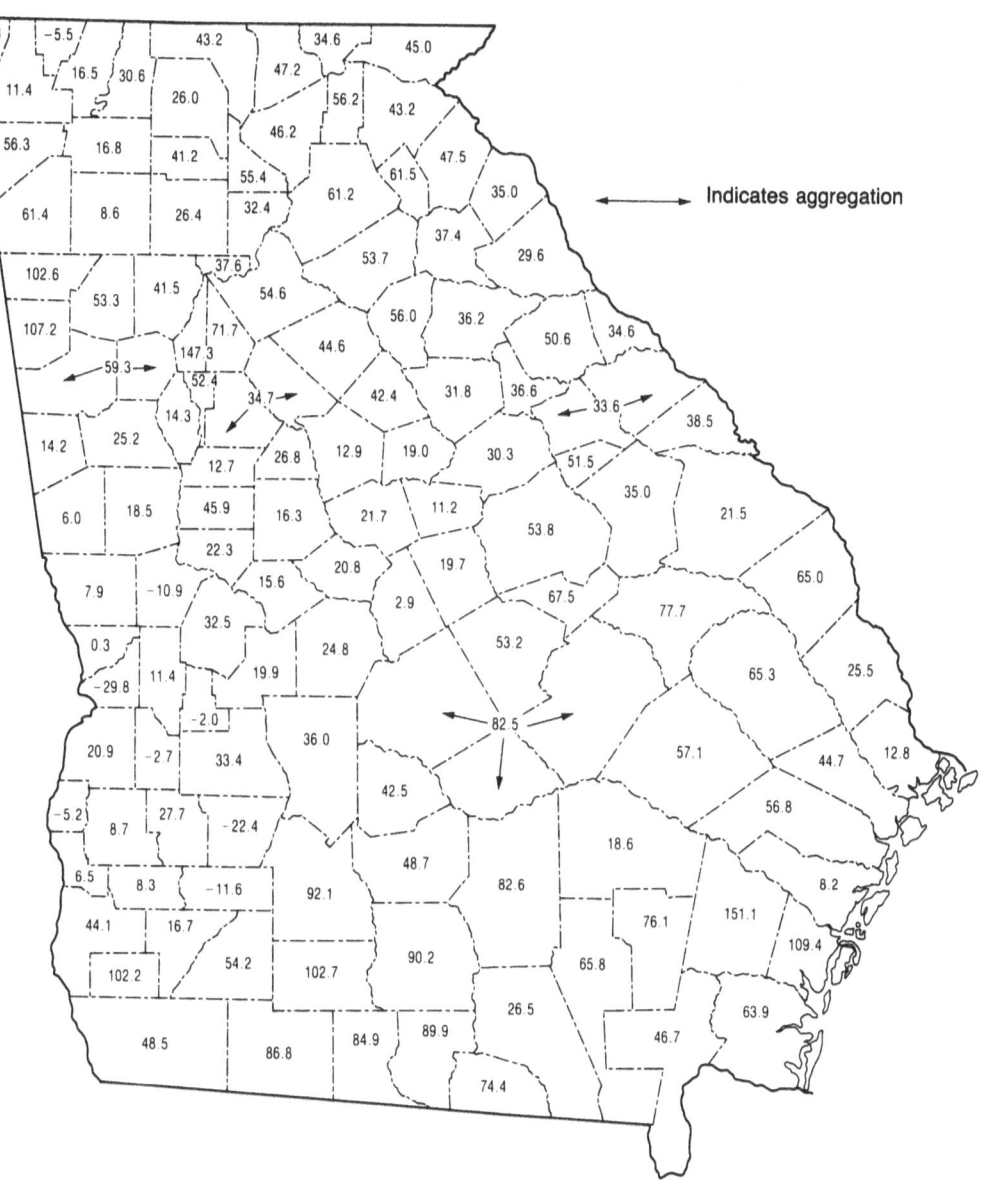

Notes: The state mean is 38.1 percent. Because county boundaries changed significantly after the war in some regions of Georgia, four groups of 1860 counties have been aggregated to achieve comparability of areal units for the two years.
Sources: U.S. Census Office, Eighth Census [1860] and Tenth Census [1880]. *Population of the United States* (Washington, D.C.: GPO, 1864, 1883).

cause of boundary changes that occurred following the war. Such percentage changes are not very useful in themselves. They become more useful when measured against the state mean, which in this case was 38.1 percent, thus enabling us to rank counties in terms of relative loss and gain. But percentage gains must also be compared to an estimated natural growth rate, since absolute gains may mask real losses through emigration. A natural growth rate is an extraordinarily difficult statistic to derive, even when suitable documentation is at hand. There is no reason to assume—and considerable reason not to do so—that rates of natural growth will be constant throughout the state. Maturation and age of marriage, and therefore birth rates, as well as death rates, can be expected to vary with levels of nutrition, climate, and availability of land. The list could be extended. We shall frankly ignore these difficulties and engage in an analysis that lacks such refined statistics and offers only crude generalizations. We shall employ the actual white growth rate for all southern states as a surrogate for a natural growth rate. Because of the rate's unreliability, only large departures from it should be considered significant.[9]

Map 7.5 presents white population percentages at four levels of relative gain and loss. Since the state mean white population increase was 38.1 percent, and our natural growth estimate should have produced a 49.7 percent increase (which was the increase in white population in all southern states between 1860 and 1880), it is credible that Georgia suffered a substantial loss in white population as a result of emigration. Level I, which will represent the primary zone for white immigration, is therefore defined by those counties that experienced a growth in white population in excess of 49.7 percent. Level II, defined by those counties that exceeded the state mean of 38.1 percent but fell short of 49.7 percent, represents a secondary zone of population movement characterized by emigration, undoubtedly partly offset by immigration; these counties suffered a loss of white population but at lower levels than experienced by the state as a whole. Level III comprises a zone defined by those counties that achieved absolute gains in white population but fell below the state mean. Counties falling into level 4 suffered absolute losses of white population. There is no clear demarcation at zero growth from an analytic perspective, and those counties that are marginally above zero should especially be noted.

Map 7.5. Percentage of Change in White Population (Shaded), 1860–1880

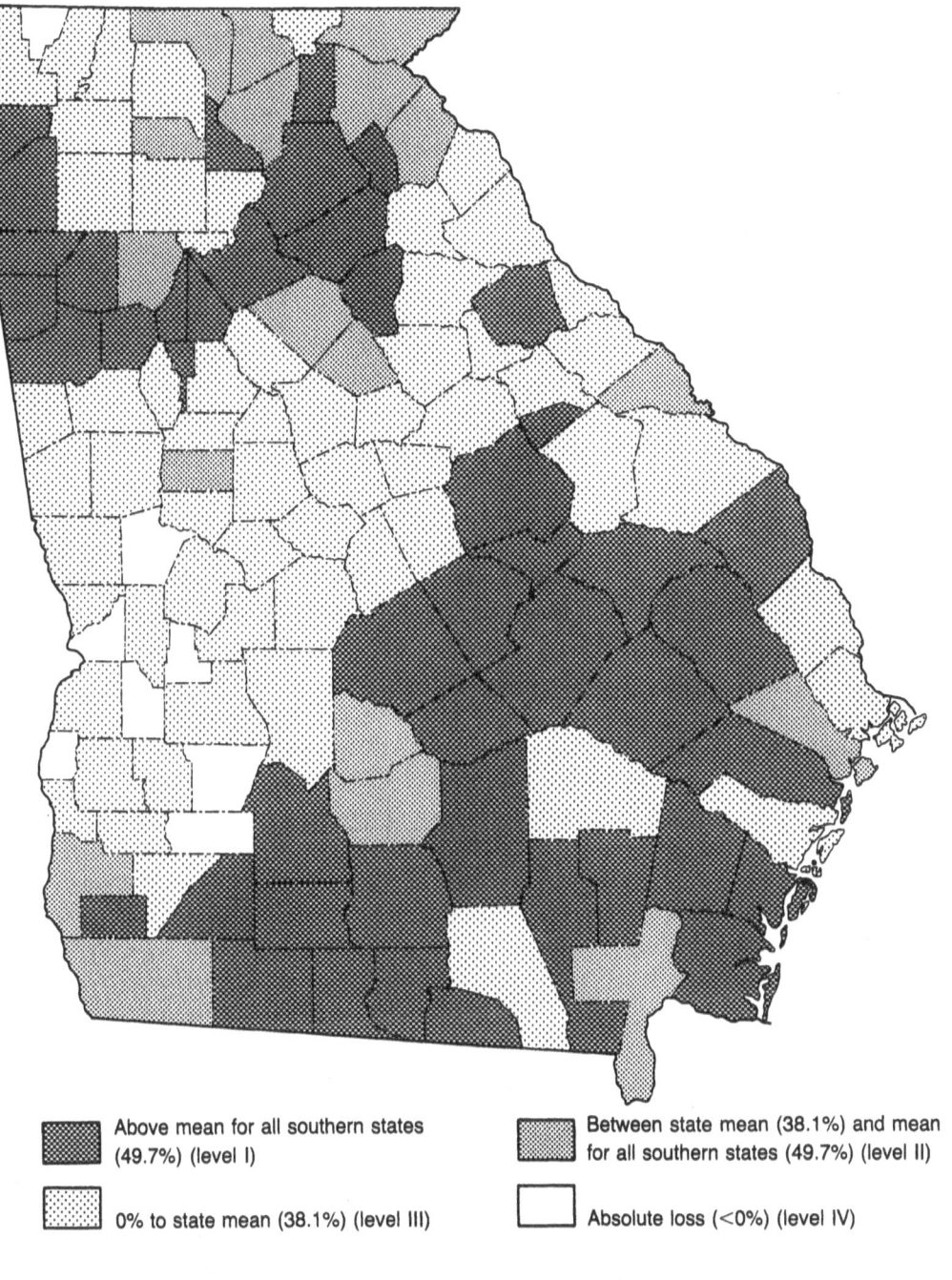

Above mean for all southern states (49.7%) (level I)

Between state mean (38.1%) and mean for all southern states (49.7%) (level II)

0% to state mean (38.1%) (level III)

Absolute loss (<0%) (level IV)

Note: Because county boundaries changed significantly after the war in some regions of Georgia, four groups of 1860 counties have been aggregated to achieve comparability of areal units for the two years.
Sources: U.S. Census Office, Eighth Census [1860] and Tenth Census [1880]. *Population of the United States* (Washington, D.C.: GPO, 1864, 1883).

Patterns of white population movement emerge on this map with unusual clarity. Counties in the upper Piedmont and in the more productive areas of the lower Ridge and Valley, along with the Pine Barrens and Limesink, experienced postwar growths in white population that far exceeded both the state mean and natural growth. On the other hand, the less productive of the mountain counties and the lower Piedmont were suffering relative losses in white population, the losses on the lower Piedmont being greater as one proceeds westward onto what had been the newer and richer lands. The most catastrophic losses of white population occurred in the areas that had been in the forefront of the cotton economy before the war, that is, in the counties of southwestern Georgia, both above and below the fall line. Whites in Georgia were moving to the fresher agricultural lands toward the periphery of the old Cotton South, down out of the more remote mountains, and, except for Fulton County (Atlanta), out of the urban counties of Chatham (Savannah), Richmond (Augusta), Baldwin (Milledgeville), Bibb (Macon), and Muscogee (Columbus).

Percentage gains and losses of black population during the same period are presented on map 7.6. Map 7.7 again classifies these percentages into four zones of population movement, but the levels defined are different from those for whites. The state mean for black population growth was 55.7 percent, which is substantially above the mean growth rate for blacks in all southern states (44.4 percent). Unlike whites, who were moving out of the state in large numbers, blacks were entering Georgia in equally large numbers following the war. Level I, constituting the primary zone of black immigration, is therefore defined by those counties that fell above the state mean of 55.7 percent. Level II, which is a secondary zone of immigration and (in contrast with this level for whites) one of real growth, is defined by counties falling above the black growth rate for all southern states (44.4 percent) but below the state mean of 55.7 percent. Level III contains those counties that experienced absolute growth but at less than the mean for all southern states and therefore probably, given the roughness of the estimate, suffered relative black population loss. Level IV is composed of those counties that suffered an absolute loss.

The pattern that emerges on the map is not so clear as that for whites, partly because of the enormous inflows of blacks following the war. But the redistributions are nevertheless striking. Given the usual anomalies, such as Gwinnett, blacks were moving into precisely the

Map 7.6. Percentage of Change in Black Population (Statistics), 1860–1880

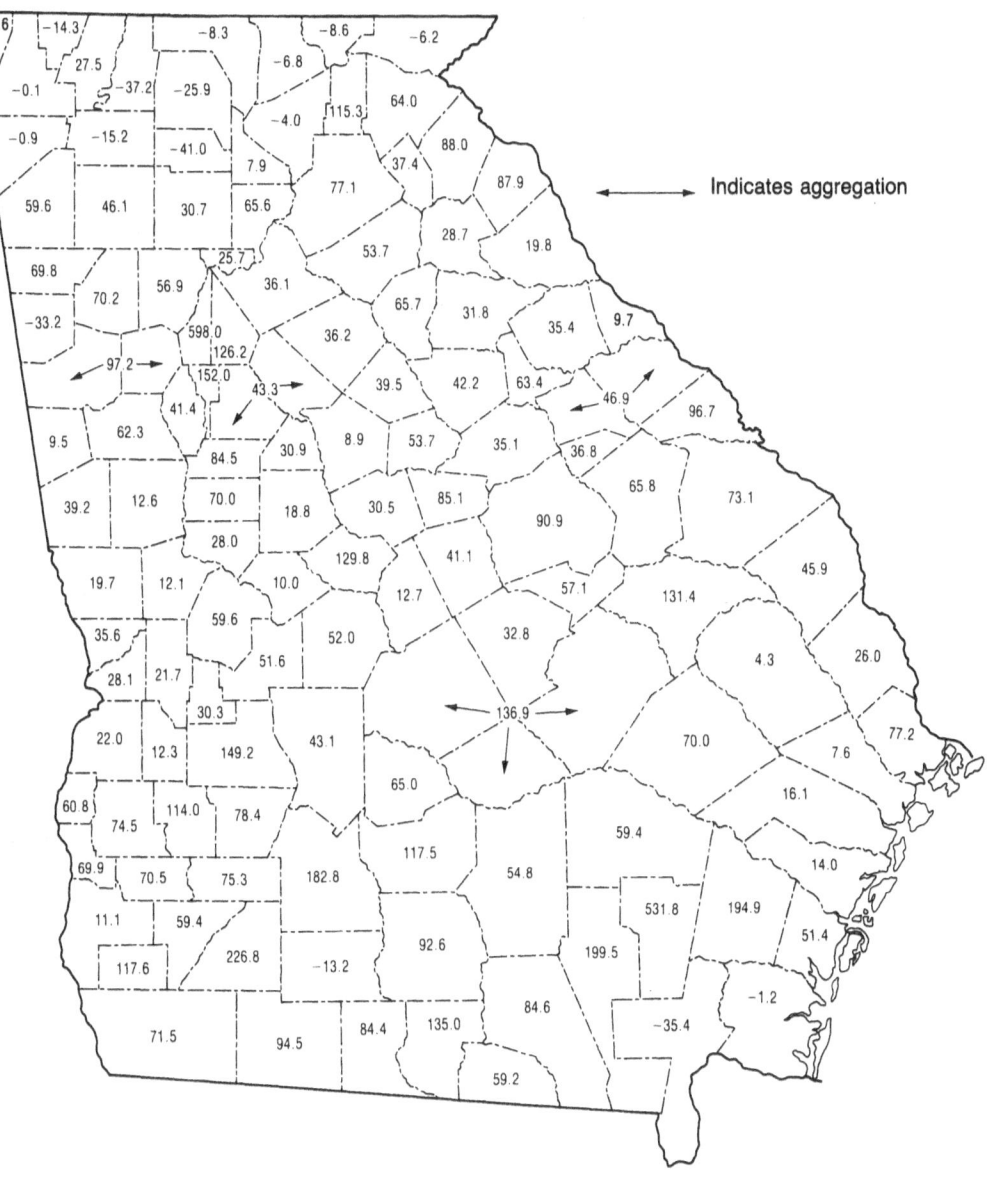

Notes: The state mean is 55.7 percent. Because county boundaries changed significantly after the war in some regions of Georgia, four groups of 1860 counties have been aggregated to achieve comparability of areal units for the two years.
Sources: U.S. Census Office, Eighth Census [1860] and Tenth Census [1880]. *Population of the United States* (Washington, D.C.: GPO, 1864, 1883).

Map 7.7. Percentage of Change in Black Population (Shaded), 1860–1880

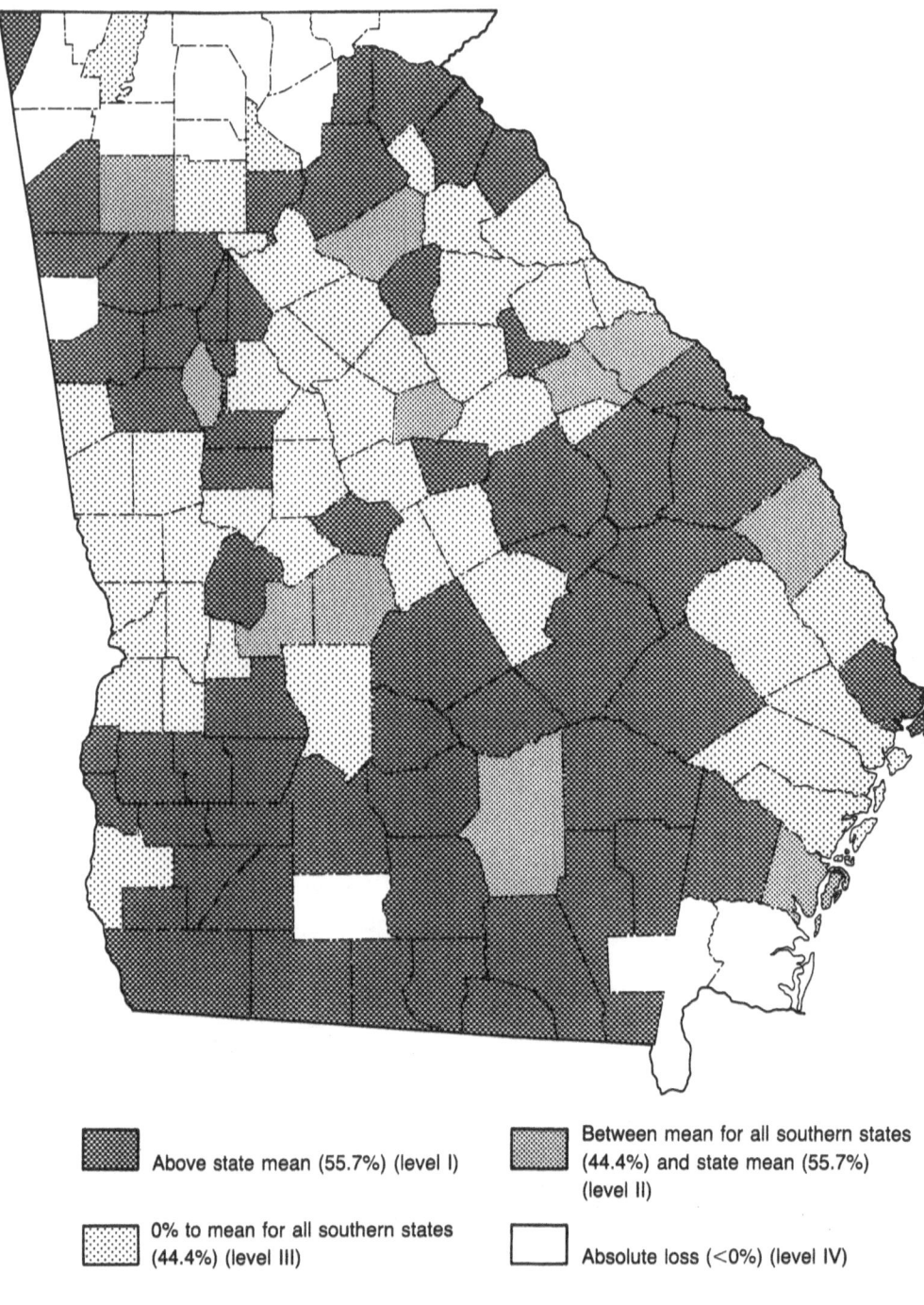

Above state mean (55.7%) (level I)

Between mean for all southern states (44.4%) and state mean (55.7%) (level II)

0% to mean for all southern states (44.4%) (level III)

Absolute loss (<0%) (level IV)

Note: Because county boundaries changed significantly after the war in some regions of Georgia, four groups of 1860 counties have been aggregated to achieve comparability of areal units for the two years.
Sources: U.S. Census Office, Eighth Census [1860] and Tenth Census [1880]. *Population of the United States* (Washington, D.C.: GPO, 1864, 1883).

same areas as whites, with one important exception: the area of most significant white loss, the southwestern counties below the fall line, was the area of most significant black increase. Black population increase within the old black belt was not insignificant, however. Maps 7.8 and 7.9, which categorize counties by their percentage of slave population in 1860 and black population in 1880, show that, while the outer spatial perimeters of the black belt were not substantially enlarging, the levels of black population within regions of highest staple production before the war were greatly increasing. Apart from the older counties of the eastern Piedmont, where relative white population stability tended to stabilize population ratios between whites and blacks, and the insalubrious coastal counties, where the staple economy was in sharp decline with the loss of slave labor, black population ratios rose sharply throughout the old areas of prime staple production. Nevertheless, with the important exception of the southwest, black population increase, when judged by magnitudes, was most significant in those counties on the margins of the old staple economy and into which whites were also moving. While negative growth rates occurred in the mountains and in the Okefenokee swamp, the movement was not significant, since the numbers of blacks in such counties before the war were small.

Map 7.10 presents percentage increases in improved acres between 1860 and 1880 for each county. Interesting patterns are revealed. If Appling and Bulloch counties are discounted as outliers, the Pine Barrens appear to have been engaged in substantial clearance, with improved acreage increasing moderately in some counties and sharply in others. As we have seen, the importance of tenancy was diminishing in this region, and the absolute number of proprietors was doubling. In the counties of antebellum staple production, where tenancy rates were rising steeply, improved acreage was markedly down. Only in the upper Piedmont and lower mountain counties were tenancy rates and improved acreage simultaneously increasing, but in these regions proprietorship was increasing substantially as well. It appears from the map, then, that by 1880 tenancy was negatively associated with clearance, except perhaps in the northern regions of new staple production.[10] But such a conclusion can be quite misleading, since it is doubtful that the data can sustain any reliable conclusions regarding the relations between tenancy and clearance. The unreliability of the data arises from their failure to delineate patterns of clearance that un-

Map 7.8. Slaves as Percentage of Total Population, 1860

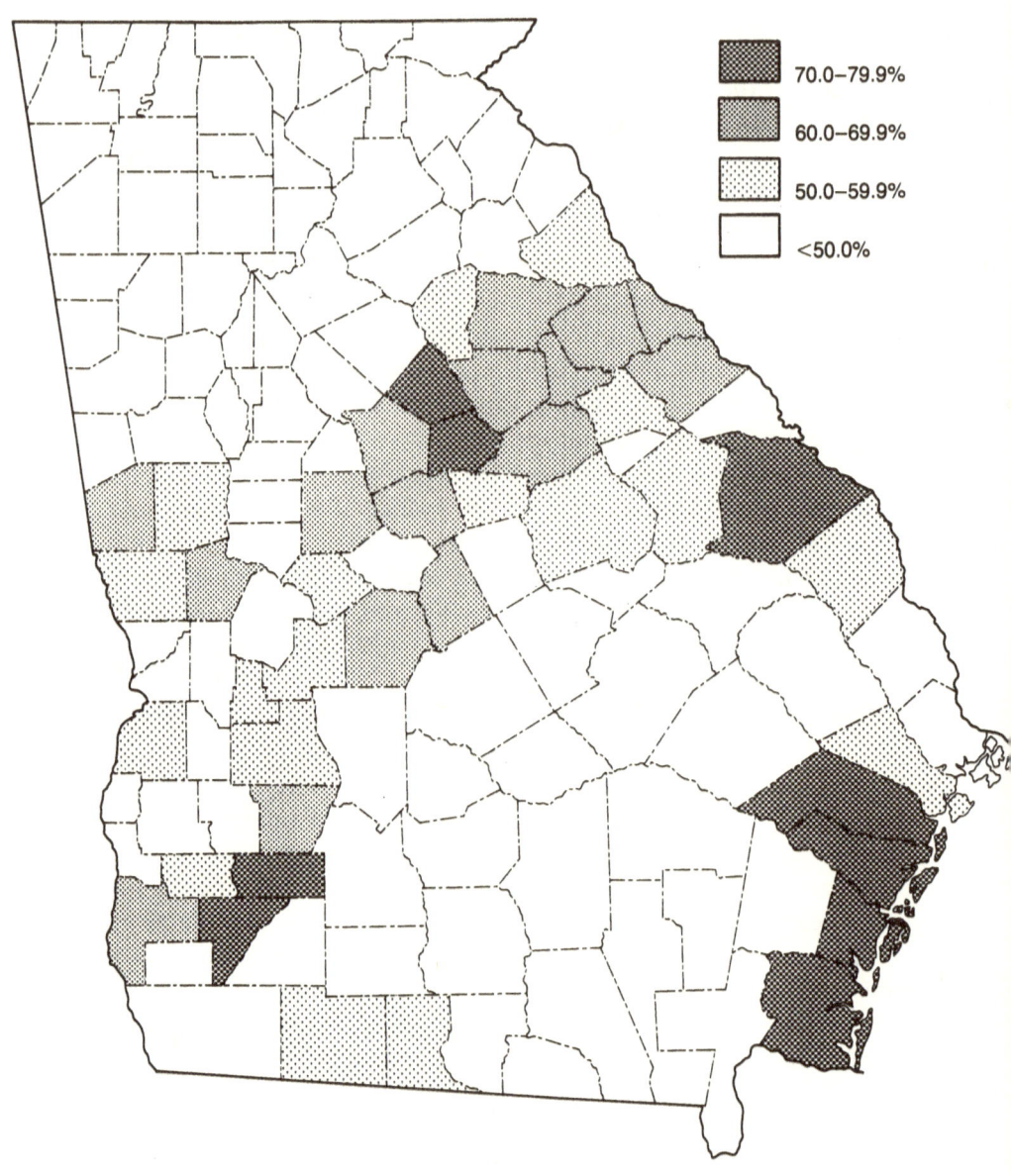

Note: The state mean is 43.7 percent.
Source: U.S. Census Office, Eighth Census [1860]. *Population of the United States* (Washington, D.C.: GPO, 1864).

Map 7.9. Blacks as Percentage of Total Population, 1880

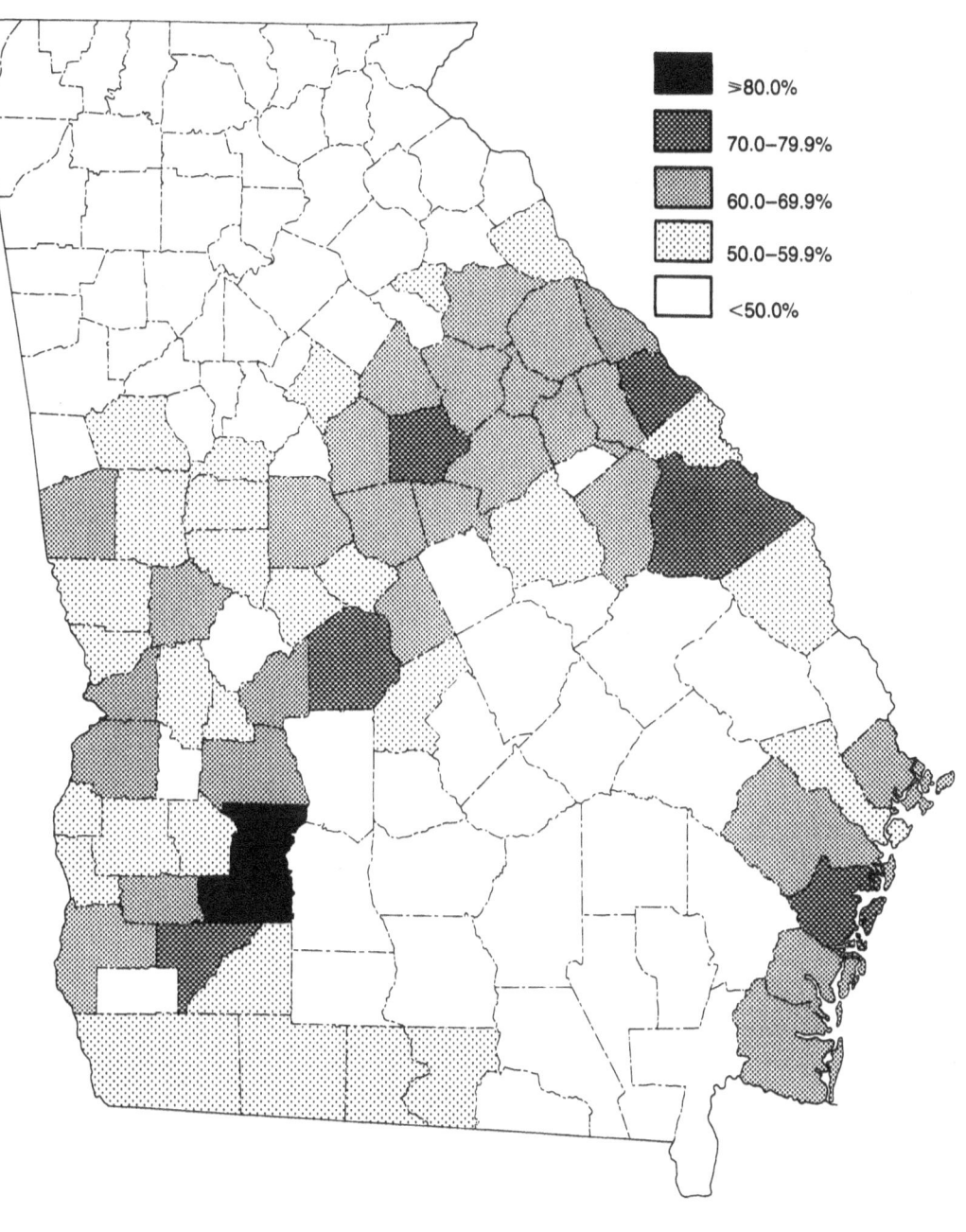

Notes: The state mean is 47.0 percent. See 1880 map on page xxi.
Source: U.S. Census Office, Tenth Census [1880]. *Population of the United States* (Washington, D.C.: GPO, 1883).

Map 7.10. Percentage of Change in Improved Acres, 1860–1880

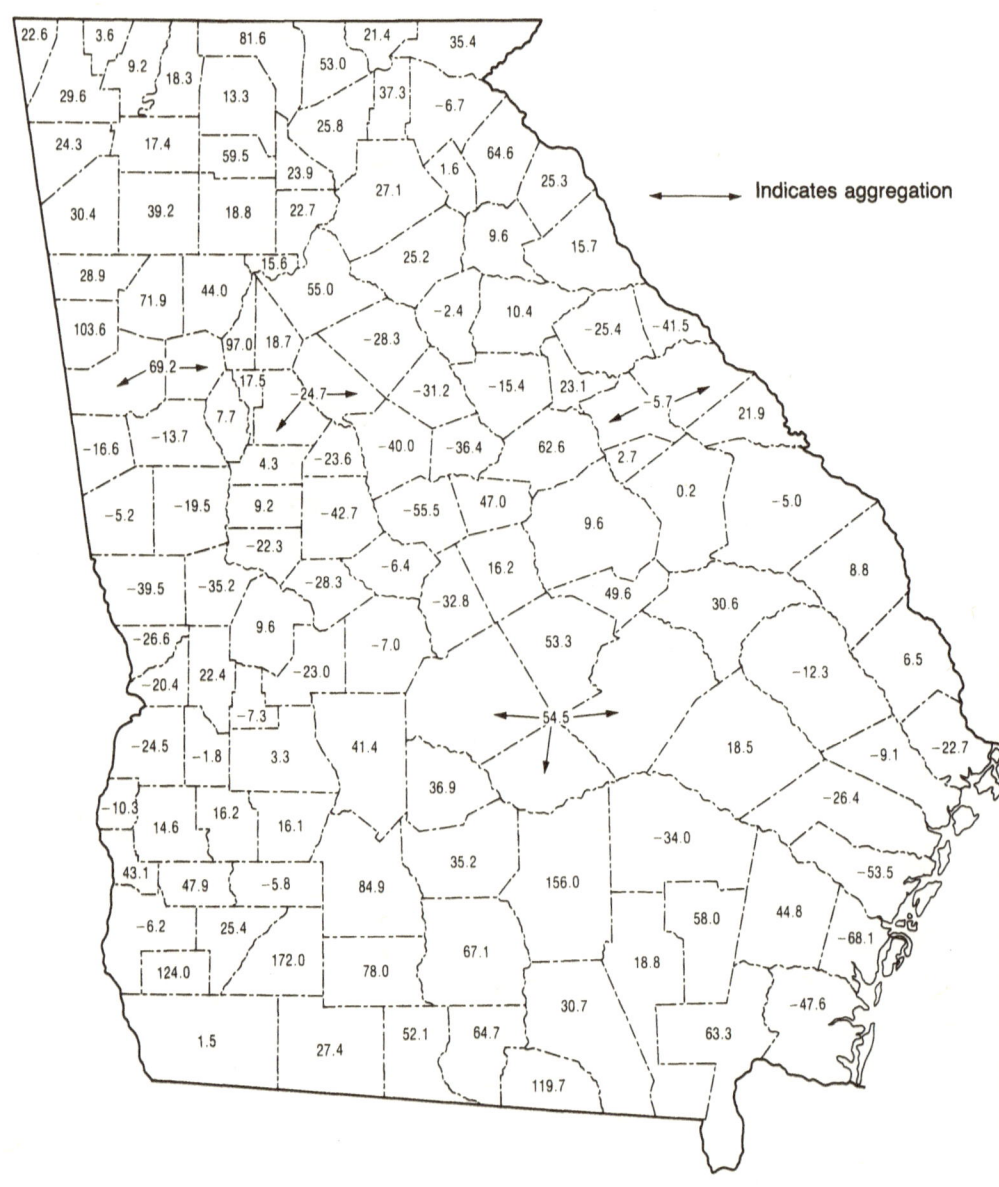

Notes: The state mean is 1.8 percent. Because county boundaries changed significantly after the war in some regions of Georgia, four groups of 1860 counties have been aggregated to achieve comparability of areal units for the two years.
Sources: U.S. Census Office, Eighth Census [1860] and Tenth Census [1880]. *Agriculture of the United States* (Washington, D.C.: GPO, 1864, 1883).

doubtedly occurred within each county. Over the course of two decades some lands in each county would be abandoned, some unimproved lands would be cleared and brought into new cultivation, and—especially after the war—some lands that had been lying out would be reclaimed. It may be that such a process of internal redistribution of acres occurred in Thomas County, where large increases are reported among both proprietors and tenants but only a small increase in improved acreage is indicated. Worth County, which also reports increases in proprietors and tenants, but at half the level of Thomas, increased its improved acreage by three times the rate. It is entirely possible that counties that underwent extensive redistributions of acreage would report only marginal net gains and losses. But even if the data were more reliable, it would be wrong to superimpose an antebellum explanation of tenancy distributions on postbellum agriculture, even if many of the same variables are involved. Our association of clearance with tenancy before the war, to the extent it holds true, assumes a rational choice between the application of slave labor and the various forms of free tenancy, the former being in short supply and too expensive to waste on clearance. After the war, slave labor no longer formed part of the equation, and the choice lay between forms of tenancy.

A more promising explanation of tenancy redistributions following the war is the obvious one: tenancy, along with population, reacted directly to altering spatial patterns of staple production. Map 7.11 displays percentage increases in bales of cotton between 1859 and 1879. Because the average 1880 bale was supposed to weigh 475 pounds in lint cotton, and the average 1860 bale purportedly weighed 400 pounds, all percentages on this map should be adjusted upward by 18.8; thus, if we accept these statistics at face value, no percentage increase may be considered a real gain unless it exceeds −19 percent. However, any reliance on these statistics should be tempered by a perusal of the following comments by R. H. Loughridge, who wrote the Georgia section of the 1880 Cotton Report.

> This weight of the bale [475 pounds] is the average reported for the season from Savannah and Atlanta, an increase over that of 1870, due probably to the more general use of improved presses having a greater power and capacity for heavier bales, and also because transportation charges are *per bale,* irrespective of actual weight.... The rates of transportation over the different routes vary according to the relation of com-

Map 7.11. Percentage of Change in Number of Bales of Cotton, 1859–1879

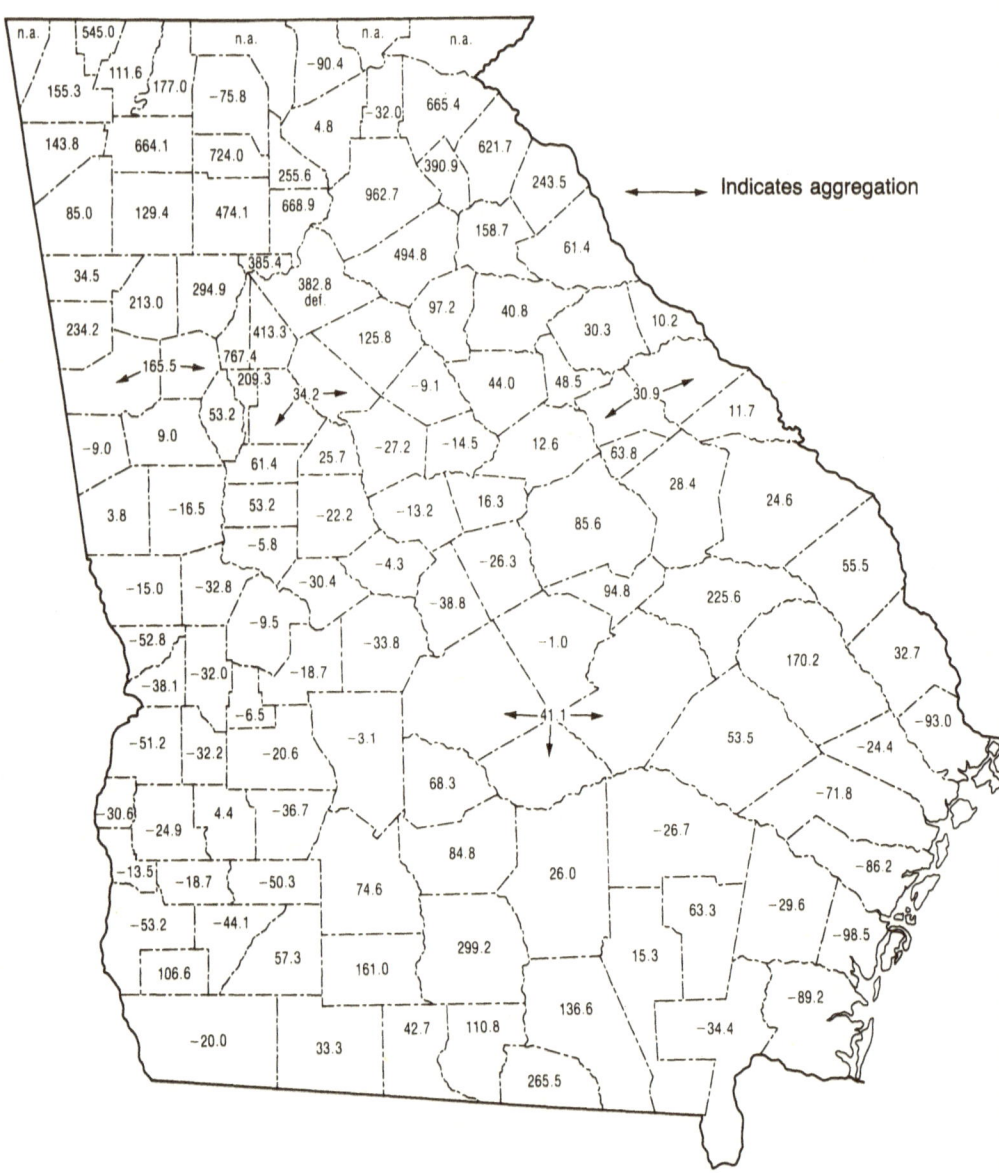

Notes: The state mean is 16 percent. Because county boundaries changed significantly after the war in some regions of Georgia, four groups of 1860 counties have been aggregated to achieve comparability of areal units for the two years.
Sources: U.S. Census Office, Eighth Census [1860] and Tenth Census [1880]. *Agriculture of the United States* (Washington, D.C.: GPO, 1864, 1883).

peting lines. The rates were once fixed *per bale of any weight* above 300 pounds, and it was therefore to the advantage of the shipper that the weight of a bale should be as great as possible, limited, of course, by other expenses, such as the extra handling and drayage incident to such an increased weight. These weights frequently reached 600 pounds, but 475 pounds was in 1879 the usual bale. The irregularity of charges made by the railroads caused such a feeling of dissatisfaction in the state that a railroad commission has been created by the legislature and a schedule of rates as uniform as possible has been adopted, not by the bale, as heretofore, but by the pound, and the tendency now is a reduction in the weight of the bale for greater convenience in handling. This commission has just been established, and this fact will explain the discrepancy between the answers of the various correspondents in the county descriptions, which were made at different times. The ordinary bales coming from the county presses are large and bulky, and take up so much unnecessary space that transportation companies now almost invariably have them reduced from 40 to 50 percent in size before shipment by means of steam hydraulic compresses. The pressure applied varies in the different presses, the highest being about 3,800 tons per bale. So perfect are the details of manipulation that from 100 to 150 bales per hour are passed through each press. The cities of Augusta, Athens, Atlanta, Savannah, and Brunswick are supplied with these presses, all owned by private companies, the charges, which range from 25 to 60 cents per bale, being paid by the transportation companies.[11]

These remarks are sobering. They are potentially devastating to any precise reliance on statistics for cotton output at a micro level. The new super compressors to which Loughridge referred were new to the scene and were located only in the few major urban transhipment centers. They were physically remote from most cotton producers, and the expense of their use was entirely borne by the transport companies. If Loughridge was correct, the original bales sent to the transport companies in the cities varied widely in the range of their weights, "frequently" reaching 600 pounds[12] and "averaging" 475 pounds in Savannah and Atlanta in 1879. Such an average suggests that a large proportion of bales also were well below 475 pounds, perhaps frequently reaching only 300 pounds, which was the lower limit of the old railway rate. The range of variation for most bales could thus be as much as 100 percent or more. James C. Bonner has noted that as early as 1845 the Savannah Chamber of Commerce was urging planters to standardize their bales at between 450 and 500 pounds, which could mean that there was some tendency toward the 475-pound bale even in 1860.[13] But Loughridge makes it plain that growers had no real economic incentives to standardize their bales before the introduction of the new railroad rates and that at least until the time of his writing bales were not being standardized by growers—on the contrary.

If actual pressed bales are being reported as such on the enumerators' returns, then it is reasonable to assume—or seems more reasonable to us—that the bales reported were those pressed in the counties by the growers, not those re-pressed in major urban centers by the transport companies. On the other hand, it does not seem reasonable to us to assume that all growers, knowing the number of pounds for which they were finally paid, then uniformly proceeded to divide such pounds by four hundred in order to report true census bales to that meddlesome fellow who came knocking at their doors or following them into the fields. Some enumerators may have asked for pounds and performed the calculation themselves; some growers may similarly have performed such calculations when asked. But to simply assume that all enumerators and/or all growers performed such calculations assumes a degree of extra effort and uniformity in procedure that runs counter to the full weight of our experience with the census.

There is some direct evidence that a small number of enumerators did attempt to convert cotton into standardized bales of 400 pounds, as requested by the census instructions. In Habersham County, as can be seen in the documentation to appendix G, the enumerator reported the bales of many farms in fractional values. A single fractional value was also reported in Worth. No other counties in our sixteen-county sample reported fractional values, although all these counties included some number of small producers reporting as little as one bale. In applying the short method to the remaining counties of the state, we noted an additional small number of counties who reported fractional bales. Those counties that reported fractional bales generally reported them throughout the return, a few occurring on each page. There is no reason to assume that in all other counties cotton was produced in amounts evenly divisible by four hundred. One of three things must be true: either the enumerators in most counties were rounding their recalculated bales; or there was no reason to round, because the bales being reported were the number shipped—or both. While the evidence is hardly conclusive, it suggests that no easy assumptions can be made regarding the meaning of "bales" in the returns.

It may be, of course, that variations in bale weight averaged out and that the importance of error is not significant for some types of analysis. The mean weight may be sufficient. But it is equally possible that some forms of analysis will be significantly affected because variations were not well distributed. Much depends on where the various county presses were located. The 1880 report suggests indirectly that they may have been located in accordance with transport considerations, but we do not know that this is so. It may also be that large proprietors and managed plantations were more likely to recalculate into census bales; we are ignorant here as well. It is clear, however, that both error and bias may exist in the reporting of this variable, and to an unknown degree.

It is therefore not clear whether an upward adjustment of 18.8 ought to be made to the percentages presented in map 7.11. Fortunately, the questions being posed in this chapter require only very large differences in percentages, and such differences do conveniently appear on the maps. The patterns are so striking that there can be no doubt about what happened geographically to Georgia cotton production following the Civil War. The counties of the upper Piedmont constitute an abrupt ledge of increased cotton output, with a few transitional counties falling just below Atlanta. Tenancy rates for 1880 (map 7.1) form the same ridge along an almost identical line except that the tenancy values above the line are lower than tenancy values below it. Map 7.12, which gives yields of lint cotton per acre in 1879, again displays the same line of demarcation. Counties above that line, in the upper Piedmont and in the southwestern portion of the Ridge and Valley, were by now the most intensive cotton producing regions of the state. Counties in the lower Ridge and Valley had already been increasing their production before the Civil War, perhaps under the stimulus of improved rail transport. But the introduction of large-scale commercial fertilization during the late 1870s, which made it possible to accelerate plant maturation and shorten the growing season, released the region from many of the restrictions imposed on it by its altitude. The percentages of improved acres, here and in the upper Piedmont, increased, and so did the absolute number of proprietors. Tenancy rates also increased, but they did not spiral as they did in the more southerly counties of the lower Piedmont, with their older lands and declining production.[14]

While cotton production declined on the lower Piedmont, it dropped catastrophically in large numbers of counties in the Central

Map 7.12. Pounds of Lint Cotton per Acre under Cotton, 1879

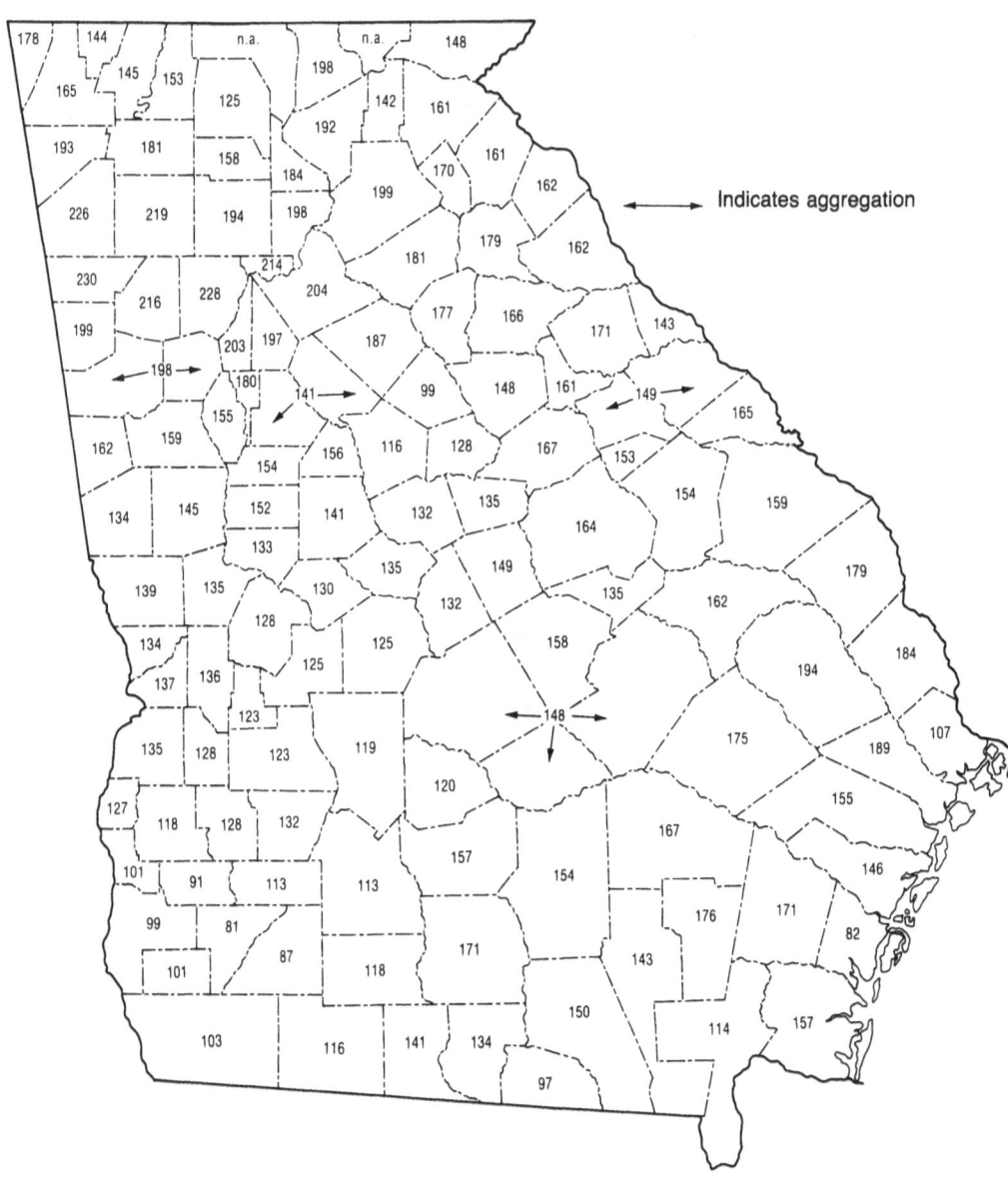

Notes: The state mean is 148 pounds per acre. Four groups of counties affected by post-1860 boundary changes have been aggregated to facilitate comparisons with maps 7.10 and 7.11.
Source: U.S. Census Office, Tenth Census [1880]. *Report on Cotton Production in the United States* by Eugene W. Hilgard et al. (Washington, D.C.: GPO, 1884), Georgia section, pages 3–5.

Cotton Belt, especially in those southwestern counties that had been the chief cotton producing centers on the eve of the Civil War. The yields of lint cotton per acre indicate that these once highly productive lands had been worn out. White farmers were pouring out of these southwestern counties. Blacks were pouring into them in far greater proportions, and by 1880 Dougherty and Lee were over 80 percent black, the highest ratios in the state and higher than any ratios achieved in 1860. Dougherty also attained the state's highest tenancy rate, with Baker a strong second. While it is true that slave ratios were already unusually high in this subregion before the war, it nevertheless remains that an additional large inflow of black population had entered the area by 1880.

Cotton production made gains in the Limesink division and in the Pine Barrens, but percentage gains in output can be misleading. The percent of tilled acres under cotton (map 7.13) gives a truer picture of the relationship of these counties to the cotton economy in 1879. The Limesink, unlike 1860, was now only approaching the state mean. Cotton would not invade the Pine Barrens on a large scale until the turn of the century, accompanied then by a new wave of immigration. Despite their relative dissociation from the cotton economy, the Limesink and Pine Barrens, along with the Palmetto Flats, were prime areas of immigration in the 1870s for both whites and blacks. Unlike in other regions of the state, this heavy inward shift of population seems to have been overwhelmingly toward proprietorship and either marginally or not at all toward tenancy. Tenancy rates, especially in the southwestern section of the Pine Barrens, dropped sharply, while the number of proprietors doubled everywhere.

Some evidence for the distribution of white tenancy and for the black share of tenancy in 1880 may be gleaned from the sample of the Cotton South collected by Ransom and Sutch (table 7.2). It is once again apparent that movements in white tenancy rates were not uniformly distributed throughout the state. If Gwinnett may be taken to be representative of the newer cotton lands of the upper Piedmont, it is clear that there were no sharp rises in white tenancy in this region. White tenancy rates had risen perhaps five or six percentage points since 1860, but there is certainly no evidence of a sharp "fall" into tenancy among whites in this region. Moreover, relatively few blacks in this county were able to gain a holding, either as proprietors or as tenants. Prime cotton lands remained largely a white preserve. On the

Map 7.13. Percentage of Tilled Acres under Cotton, 1879

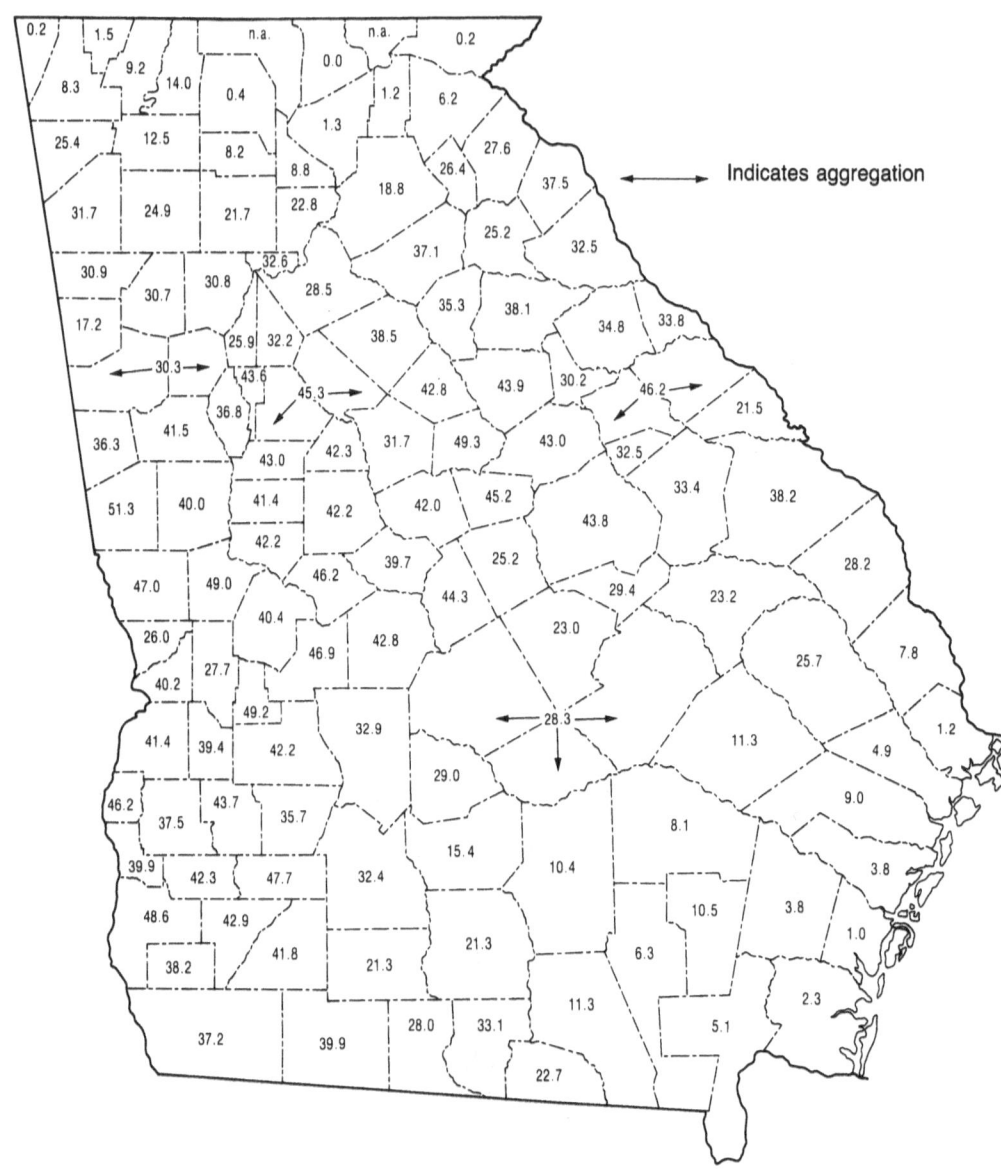

Notes: The state mean is 34.0 percent. Four groups of counties affected by post-1860 boundary changes have been aggregated to facilitate comparisons with maps 7.10 and 7.11.
Source: U.S. Census Office, Tenth Census [1880]. *Report on Cotton Production in the United States* by Eugene W. Hilgard et al. (Washington, D.C.: GPO, 1884), Georgia section, pages 3–5.

Table 7.2

TENANCY RATES (%) BY RACE, NINE COUNTIES, 1880

County	Race	Sample Farms	Renter (%)	Share (%)	Sample Total (%)	Census Total (%)	Sample Tenants
Coweta	White	149	10.7	30.9	41.6		62
	Black	79	22.8	65.8	88.6		70
	Total				57.0	61.3	
Gwinnett	White	126	0.8	37.3	38.1		48
	Black	13	0	61.5	61.5		8
	Total				40.7	42.1	
Liberty	White	46	6.5	0	6.5		3
	Black	86	53.5	4.7	58.2		50
	Total				39.3	39.1	
Tattnall	White	81	1.2	4.9	6.2		5
	Black	17	11.8	11.8	23.5		4
	Total				9.0	9.0	
Taylor	White	66	9.1	13.6	22.7		15
	Black	10	50.0	30.0	80.0		8
	Total				31.3	38.0	
Terrell	White	70	5.7	14.3	20.0		14
	Black	11	36.4	36.4	72.7		8
	Total				26.2	32.2	
Thomas	White	56	1.8	12.5	14.3		8
	Black	28	7.1	35.7	42.8		10
	Total				25.5	38.1	
Twiggs	White	37	16.2	5.4	29.6		8
	Black	51	45.1	52.9	98.0		50
	Total				67.7	60.2	
Worth	White	60	0	25.0	25.0		15
	Black	9	0	100.0	100.0		9
	Total				34.3	30.4	

Source: Data compiled from the Ransom and Sutch machine-readable sample and kindly provided by Donghyu Yang. Dubious observations have been removed by Dr. Yang.

now older cotton lands of the western Piedmont, in Coweta County, white tenancy rates seem by contrast to have risen sharply, with blacks moving into tenancy even more heavily and perhaps outnumbering white tenants. In the fall line and Sand Hills counties white tenancy rates seem to have remained curiously stable since 1860, as in Taylor, or to have dropped sharply in favor of a huge rise in black tenancy, as in Twiggs. Twiggs may in fact be characteristic of tenancy shifts occurring on the older lands of the eastern lower Piedmont and fall line. Terrell County, in the upper southwest, repeats the white pattern of

Taylor. If whites were pouring out of the cotton lands of the upper southwest and lower western Piedmont, those who remained were engaged in tenancy at about the same rates as their predecessors in 1860. On what had once been the prime lands of the lower southwest, however, white tenancy rates had dropped sharply by 50 percent or more, and blacks made up the shortfall. Worth County, in the Limesink, seems to have maintained tenancy rates that were almost precisely those of 1860, with whites giving way only marginally to blacks. Tattnall County in the Wiregrass remained, as always, largely unaffected by tenancy. The coastal rice counties, which had little or no tenancy before the war, experienced a collapse of staple production and a sharp rise in tenancy, almost all of it black. In short, if whites fell into tenancy, they did so only on the oldest lands of the eastern Piedmont and Central Cotton Belt and, perhaps, on the relatively old lands of the western Piedmont around Coweta County—if the Ransom and Sutch sample is reliable for such conclusions.

But there is evidence that the sample is not sensitive to variations in tenancy, however reliable it might be for other variables. The persistent shortfalls and overages in county tenancy rates, when compared with the aggregate census, clearly indicate that the sample is not representative with respect to tenancy, especially in a county such as Thomas; and the small numbers of blacks included in most counties, given their numbers in the general population, suggest that they may have been undercounted. In short, while data from the Ransom and Sutch sample support the conclusions presented in the preceding paragraph, they should be treated with some diffidence until the nature of the sample with respect to tenancy has been examined more closely.

The conclusions one reaches concerning the relationship between tenancy, population movements, and shifts in staple production are not very surprising ones. White population remaining in the state tended to move in two directions following the war. One movement was from the high mountains and older cotton lands to the fresher lands of the upper Piedmont and southwestern Ridge and Valley. This movement is clearly associated with shifts in cotton production. In the context of this movement, tenancy tended to rise most dramatically in those subregions where the soil was most exhausted for staple production, with white emigration being heaviest in those declining counties. Blacks, who also moved in the same two directions, moved disproportionately onto those poorest soils that experienced large losses of white population, but moved heavily as well into the newer and richer lands of northern Georgia. On balance, they seem to have done less well than whites in competing for the better tenancies, but firm conclusions here must be left to studies at the micro level, which are able to take explicit account of black and white distributions within specific counties. A second movement of population was directed into the Limesink and Pine Barrens of southeastern Georgia. This population movement may have been associated with declining cotton productivity in the Central Cotton Belt and lower Piedmont, where white populations generally declined, but we have no direct evidence for such a relationship. It is certain only that there was no alteration in staple production in the southeast of sufficient significance to attract high levels of new population. The opportunities for new proprietorship seem instead to have been the incentives. And since tenancy rates fell in these southeastern regions, it is tempting to push on to one additional conclusion: that tenancy in the late 1870s was associated with staple production and that the association was strongest where the soils were the most depleted.

8

Conclusion

THIS STUDY HAS PRESENTED two separable but interdependent sets of conclusions. The first set concerns patterns of tenancy in late antebellum Georgia and their relationship to postbellum patterns. The second set of conclusions concerns the interpretation and reliability of the 1860 manuscript census as a data source. Our evaluation of the census has been guided by two fundamental principles. First, all types of record entries and every variation in enumerator procedure must be classified and accounted for—without any exception—and some provisional basis must be laid for systematically distinguishing between intelligible procedural variations on the one hand and genuinely defective records on the other. The basis for such decision-making must be laid out in a clearly articulated set of rules. Second, after systematically classifying all record types, we have then sought *minimal patterns of intelligibility* in modes of enumeration.

These procedures have led us to two basic conclusions. First, while individual county enumerators may occasionally display internally inconsistent patterns of reporting, the most significant problems arise when the various county enumerators are inconsistent among themselves. If there is any one conclusion upon which we would insist, it is that enumerators followed procedures that varied among themselves on every occasion when ambiguous instructions permitted such variation. We also believe they may have inconsistently reported some variables, such as cotton, when the instructions seem quite clear. Such variation is pervasive and therefore affects all counties (in Georgia at least), effects large alterations in some values, and should be expected in many intracounty comparisons between censal years as well as in

Conclusion   179

intercounty comparisons within a single censal year. The use of a full record-type classification scheme—either the one we have developed or an alternative—is therefore essential to understanding and standardizing the data for comparative analysis. Our hope is that our classification scheme will provide a fruitful basis for future discussion and will prove to be one of our more important contributions.

Our second conclusion is that land tenancy, when it was reported at all on schedule IV, was reported in conventional modes that were probably devised by enumerators in an *ad hoc* fashion. We are not suggesting that enumerators exercised great intelligence in completing their returns, and we hope we shall not be misunderstood on this point. While the intelligence of enumerators cannot be determined, a clear impression of their literacy and of the care they exercised is almost unavoidable. With some notable exceptions enumerators were highly illiterate and careless. Some records in any county are demonstrably defective; we have presented a set of rules that can identify such records, and we have noted those encountered in our sample in our documentation to appendix G. We also fully accept the probability—a very high one—that some apparently correct records are actually defective, and we have therefore tried whenever possible to accept only large differences as significant. Surely everyone who has spent any time with the census has reached similar conclusions.

But we have reached two further conclusions that we believe are more distinctive. First, while it is true that individual enumerators were careless and were to some degree capable of internal inconsistency, it nevertheless also remains true that they individually followed (and developed) procedures in response to their instructions that have left clearly perceptible patterns in the returns. Our case rests on the intelligibility of the returns—on the patterns to be found in them—and not on the intelligence or clever inventiveness of the enumerators. The patterns are not in fact very cleverly or intelligently devised for statistical purposes; the problems they produce for modern statistical analysis are the focus of much of our discussion. But we have tried to make the case that the patterns, or at least the dominant ones, are logically related to contemporary legal concepts and to the administrative culture of the region and are natural responses to ambiguous instructions. We have also tried to confirm the reality of the conventions by noting that in some counties standard and other conventions are accompanied by annotations of "tenant" in the acreage columns. Any argument that rejects conventions altogether must offer an alternative and satisfying account of these annotated missing-value entries, and there are a substantial number of them.

We would go further. Any argument that seeks to reject all conventions must also provide an alternative account of all entry types that we have identified as tenancy conventions. It is difficult to see how all conventions can be rejected unless the principle of intelligibility is rejected, that is, unless one regards the apparent patterns as no more than random inconsistency by individual enumerators producing massive numbers of records that are merely defective for one or more values. But if one starts down this path, it will end in the complete rejection of the census as a data source. Entries in conventional form, as can be seen from our tenancy rates, comprise 20 to 40 percent of schedule IV entries in most counties. If enumerators were that unreliable, then they can be relied on for nothing, and everything they did must be presumed worthless—especially since such obvious "errors" must comprise only the tip of the iceberg, and one whose depth cannot be measured. All historical data have their own peculiar defects, but there must be *some* reasonable limits to acceptability. On the other hand, if conventions are accepted as real (whether as defined by us or as revised by others), then enumerator inconsistency among counties must be taken very seriously indeed and a major reevaluation of the census as a data base is urgently needed.

Our second set of conclusions concerns patterns of tenancy in late antebellum Georgia and their relationship to postbellum patterns. These conclusions are wholly dependent on our analysis of the census as a data source. Our estimations of tenancy rates have been calculated at four levels in an attempt to establish a minimum and maximum range, and we have gone to this length precisely because we realize that it will be some time before any consensus can emerge—if it ever does—on how tenants of all types should be identified in the records. Those who accept conventions should be able at least to accept level I. While we have expressed a preference for some combination of level III and level IV estimates, we must again state that our preferred estimates must be considered as provisional. We hope that the arguments supporting our preference will be found to be substantive and rational, but we scarcely expect that they will instantly be embraced by the entire academic community. There are ample reasons why judgments on these conclusions should be judicious and slow in coming. For one thing, our analysis of conventional modes of reporting in the census entails a perceptual shift. One must be able to see the patterns and to assimilate their significance, and that takes time. But even more pointedly, it remains to be seen whether the arguments we have offered in support of our preferred tenancy estimates are susceptible to definitive confirmation (though some may be more easily disconfirmed). Certainly the census does not seem to contain the means for definitive confirmation or disconfirmation internally. The argument must proceed from an examination of noncensal materials such as land, tax, and plantation records with which we are largely unfamiliar, and the results will emerge only slowly.

But while precise levels of tenancy should properly remain a subject for discussion, we feel that those who are at least fundamentally in agreement with us in our interpretation of the census as a data source will also feel compelled to accept the conclusion that late antebellum tenancy rates were high and displayed coherent regional patterns, the degree of regional and subregional coherence depending on specific methods of interpretation that are more problematic. Our own preferred rates conform closely to those of Owsley, except when he employed tax records, and to Clark, but not necessarily to the rates of Owsley's other students who deviated from his methods. The findings of Owsley and Clark suggest that our Georgia patterns might be generalizable to other regions of the South, but great caution must be exercised on this point. The regions defined by Owsley are large and can be quite heterogeneous with significant subregional variations, as our own findings indicate, and the reliability of his estimates, even at their high level of generality, may be affected by the counties he included in his sample. It is also possible that some of the Owsley counties outside of Georgia were significantly contaminated by types b and c records. In short, we simply do not know what degree of reliance can be placed in the specific findings of Owsley and Clark. It is clear, however, that their findings have been neglected for too long; why they have been so neglected must remain one of the minor mysteries of modern historiography.

Those who agree with us that tenancy was significantly high and regionally distributed in the late antebellum period may not feel equally inclined to accept our explanation of tenancy as an alternative labor supply within the slave—or as some would say, the "dual"—economy. Obviously a tenant does not provide labor in precisely the same sense as does a slave or a free white laborer. We do not intend these remarks as a precise contribution to the debate on the theory of tenancy, and we have deliberately avoided entering into this complex discussion. Our intention has rather been to offer some rather loosely formulated generalizations that really constitute observations on the spatial variations of tenancy in 1860 and 1880 and the apparent relationship of those variations to other known variables. Thus we have noted that tenancy in 1860 is highest on the worst staple lands and lowest on the best; even within the regions of best staple soils, it tends to be low on the oldest and most exhausted lands and to increase markedly in newer regions of high productivity that are being freshly opened up for staple cultivation. From these observations we have leapt to the tentative conclusion that proprietors were making what appears to us to be a rational economic choice: since most forms of tenancy require relatively low levels of capital allocation when compared with slave or free white hired labor under direct management, some form of tenancy was therefore a preferred choice on those lands that were either poorly adapted to staple production and therefore to high capital returns or that required substantial new clearance. Similarly, when comparing the 1860 patterns with those of 1880, there seemed a clear correlation between shifting patterns of staple production and shifting levels of tenancy. We therefore again leapt to the conclusion that postbellum tenancy followed staple, as population also seems to have done; indeed, such shifts seem to have been part of a longer trend reaching back into the antebellum period. Correlations do not in themselves constitute relationships, however, and it remains to be seen whether these generalizations can be sustained by microlevel research.

We have drawn a picture of the southern economy that is far more complex than the one that normally emerges from the literature. It may be that many historians have somewhat always known that antebellum southern agriculture and social structure was more differentiated but, with the exception of those who have mused over the "dual economy," they have not written about it in that fashion. Moreover, some of those who have been most articulate on the nonplanter South, such as Frank Owsley, have focused almost entirely on the small to middling white yeoman farmer and the extent to which he characterized the economy and dominated the culture of the regions on the margin of the staple economy. Our findings suggest that the dual-economy thesis is itself a fundamentally misleading interpretation of conditions in the antebellum South as a whole. The extreme upland regions of Georgia were heavily characterized by a class of exceedingly poor tenant farmers, and in some counties a large class of apparently landless miners, who were dependent on a small number of large proprietors, many of whom were resident and, despite their holdings, not wealthy by lowland standards. Moreover, population densities were high in these mountain valleys. Such regions stand in sharp contrast to the Wiregrass section of the lowland plain, where tenancy and large arable holdings were relatively rare and population densities were persistently thin. It was the Wiregrass region that produced the independent and isolated backwoodsman who owned his holding and wandered freely through lonely tracts of unoccupied pine forest and swamp. Between these two extremes lay a richly variegated Georgia countryside. It is time that such complexity received more serious attention.

But once again, those who are prepared to agree with us that antebellum society was complex and urgently deserves further study may not be willing to accept what is easily the most contentious of our conclusions: that sharecropping in the postbellum sense constituted a portion of that complexity. It is entirely possible to accept our tenancy thesis, and even to grant that tenancy almost certainly existed in more than one form, without further granting that one of those forms was sharecropping. Alternatively, it is possible to grant that some small number of sharecroppers did exist in Georgia in 1860—after all, the term *cropper* was employed by enumerators in some few counties—without also granting that their numbers were large or of any real economic or social significance. We wish to concede quite candidly that the absence of any widespread contemporary literary reference to sharecropping is more than merely curious: it constitutes a powerful argument that sharecropping, if it did exist to any extent, was not economically or socially significant. But while we concede the importance of such an argument, we must also note that arguments from the silence of literary sources can never be definitive, and in this case they are peculiarly not convincing. The reason is a simple one: tenancy is equally not mentioned in contemporary literary sources. We have combed every antebellum and early postbellum issue of the *Southern Cultivator* and searched a large number of Georgia county histories in vain for any reference to any form of antebellum tenancy;

yet the manuscript census schedules are replete with the term or its equivalents. Historical research does turn up such curiosities from time to time. In any case, it is important that we not be misunderstood on one point: we are not arguing that a postbellum form of sharecropping *did* exist in late antebellum Georgia, nor even that it *must* logically have existed. We are instead presenting a circumstantial case for its existence, and for its existence on some significant scale in some regions. We find sharecropping to be the most plausible explanation for certain types of record entries, and for certain patterns of entries we have found in adjacent and highly comparable counties. The arguments we present in chapter 5 are not sufficiently convincing, even to us, when taken singly; they become more convincing only when their cumulative effect is weighed. We began this chapter by noting two principles that have informed everything we have done: that all types of record entry and every variation in enumerator procedure will be classified—and, we should add, will play some role in our analysis—and that minimal patterns of intelligibility in modes of enumeration will be sought. If these principles seem sensible, then arguments against sharecropping should prove more fruitful if they have more than negative content. Ideally they should provide alternative and more satisfying explanations of those specific record types and spatial patterns on which the argument is based. We believe our circumstantial case is at least sufficient to shift the burden of argument. While it would clearly be silly to ask that anyone prove that sharecropping or, indeed, any other institution did not exist, it is perhaps not unreasonable to require that a satisfying account be made of major portions of a major data source.

While our conclusions suggest some patterns of continuity between the antebellum and postbellum southern economy that run counter to the dispositions of some historians, it would be wrong to interpret them as denying that postbellum developments were of the most fundamental importance. Our tenants were overwhelmingly small and marginal producers, located principally outside the areas of heavy staple production. While they seem to have been intensive producers of cotton, and while they constituted collectively a substantial proportion of the farming community, they did not make a large contribution to the production of cotton or the wealth of the South. Patterns of tenancy were regionally differentiated and high before the war, and actually declined in some regions by 1880, but our estimates of antebellum tenancy for Georgia as a whole, along with Owsley's for the entire South, do not equal the general tenancy rates achieved within two decades following the war. Large and fundamental structural changes occurred following emancipation, and the principal mechanisms for those changes remain untouched by our findings: emancipation itself and the altered systems of credit that emerged in the late 1860s and 1870s. What our findings do suggest is that postbellum tenancy was not merely an *ad hoc* invention suddenly and hastily devised in the South as a response to emancipation. It had deep and substantial roots in southern society. Unfortunately the published census returns for 1880 do not distinguish between white and black tenants, so the precise impact of postbellum tenancy expansions on whites cannot be measured without more extensive work in the later manuscript schedules than we have been able to present. Although white tenancy rates rose sharply by the end of the century, our preliminary analysis has suggested that the picture of a white antebellum yeomanry falling into a dependent tenantry during the critical years of Reconstruction is fundamentally misleading. They fell instead into competitive relations with blacks for a scarce and valuable resource over which they had previously exercised a monopoly both as proprietors and as tenants. In this way their status was also altered, and the seeds of racial conflict were sown more thickly on the ground.

Appendix A
ESTIMATED NUMBER OF TENANTS BY CONVENTION FOR EACH COUNTY (SHORT METHOD), 1860

County	Class	Standard convention[a]	Standard convention with "tenant"[b]	Improved acres only	Improved acres only, with "tenant"[b]	Farm value only	Farm value only, with "tenant"[b]	Improved acres and farm value only	Improved acres and farm value only, with "tenant"[b]	Improved acres and unimproved acres only	Improved acres, unimproved acres, and farm value with "tenant"[b]	Schedule I "tenant"[c] also on Schedule IV	Schedule I "tenant"[c] not on Schedule IV
Appling	3												
Baker	2	8		41				10					
Baldwin	2				65								
Banks	2	268 (206)											
Berrien	4												
Bibb	3												
Brooks	3												

(continued)

Notes: The figures in parentheses omit tenants classed as types b and c. Those not in parentheses include tenants classed as b and c. An *X* indicates the occurrence of such cases in the designated county. A count of these cases has not been undertaken. An explanation of the short-method estimating procedure can be found in chapter 2.

[a] Improved acres, unimproved acres, and farm value missing.

[b] "Tenant" refers to such enumerator designations on schedule IV as "tenant," "renter," "rents land," "sharecropper," and "cropper."

[c] "Tenant" refers to such enumerator designations on schedule I as "tenant," "renter," "rents land," "sharecropper," and "cropper."

Appendix A (Continued)

County	Class	Standard convention[a]	Standard convention with tenant[b]	Improved acres only	Improved acres only, with "tenant"[b]	Farm value only	Farm value only, with "tenant"[b]	Improved acres and farm value only	Improved acres and farm value only, with "tenant"[b]	Improved acres and unimproved acres only	Improved acres, unimproved acres, and farm value with "tenant"[b]	Schedule I "tenant"[c] also on Schedule IV	Schedule I "tenants"[c] not on Schedule IV
Bryan	1												
Bulloch	4												
Burke	4												
Butts	3												
Calhoun	3												
Camden	3												
Campbell	1	256 (170)											
Carroll	3												
Cass	3												
Catoosa	4												
Charlton	3												
Chatham	3												

Chattahoochee	2	76 (63)		4 (3)	
Chattooga	1	77 (75)			
Cherokee	2	296 (264)			
Clarke	2	89 (33)	37 (36)		7
Clay	1		54		1
Clayton	5				
Clinch	1	16 (15)	42 (41)		8
Cobb	2				26
Coffee	1	21 (17)	83 (82)		
Colquitt	1	8			
Columbia	1		16 2		12
Coweta	4				
Crawford	2	74 (71)			
Dade	3				
Dawson	1	151 (144)		1	
Decatur	1	56 (54)			1
De Kalb	2	26	38		

(*continued*)

Appendix A (Continued)

County	Class	Standard convention[a]	Standard convention with tenant[b]	Improved acres only	Improved acres only, with "tenant"[b]	Farm value only	Farm value only, with "tenant"[b]	Improved acres and farm value only	Improved acres and farm value only, with "tenant"[b]	Improved acres and unimproved acres only	Improved acres, unimproved acres, and farm value with "tenant"[b]	Schedule I "tenant"[c] also on Schedule IV	Schedule I "tenants"[c] not on Schedule IV
Dooly	1	170											
Dougherty	3												
Early	3												
Echols	1							38				X	
Effingham	5												
Elbert	1	29 (28)	24	3	41								
Emanuel	2					77							
Fannin	1	13										X	
Fayette	1	268 (220)											
Floyd	1	270 (227)	56										
Forsyth	3												
Franklin	2	30 (24)						7					

Fulton	1	39	(31)			
Gilmer	2	2		3		72
Glascock	2	35	(33)			8 (7)
Glynn	3					
Gordon	4					
Greene	1	29		6		
Gwinnett	1	144		63		
Habersham	1	352	(293)		5	
Hall	4					
Hancock	3					
Haralson	1	96				
Harris	1	204	(172)			
Hart	1	110	(86)	6		24 (23)
Heard	4					
Henry	3					
Houston	3					
Irwin	1	7		5		

(continued)

Appendix A (Continued)

County	Class	Standard convention[a]	Standard convention with tenant[b]	Improved acres only	Improved acres only, with "tenant"[b]	Farm value only	Farm value only, with "tenant"[b]	Improved acres and farm value only	Improved acres and farm value only, with "tenant"[b]	Improved acres and unimproved acres only	Improved acres, unimproved acres, and farm value with "tenant"[b]	Schedule 1 "tenant",[c] also on Schedule IV	Schedule 1 "tenants",[c] not on Schedule IV
Jackson	1	12 (11)			28 (27)								
Jasper	2	58 (52)	36	1									
Jefferson	2	63 (62)										X	X
Johnson	1	42 (41)											
Jones	3												
Laurens	1	3				89							
Lee	1	7						21					
Liberty	3												
Lincoln	3												
Lowndes	1							55				X	
Lumpkin	1	293 (156)		38 (35)									
McIntosh	3												

Macon	3					
Madison	1	285 (142)			16	X
Marion	1	7 (6)	76			7
Meriwether	1	209 (206)				
Miller	3					
Milton	1	268				
Mitchell	4					
Monroe	5					
Montgomery	1		1		12	
Morgan	3					
Murray	4					
Muscogee	2	14 (13)	2	1	9	
Newton	1	333 (165)	86 (79)			
Oglethorpe	1	48 (45)				7
Paulding	2	30	4			X	X
Pickens	1	197				
Pierce	3					

(*continued*)

Appendix A (Continued)

County	Class	Standard convention[a]	Standard convention with tenant[b]	Improved acres only	Improved acres only, with "tenant"[b]	Farm value only	Farm value only, with "tenant"[b]	Improved acres and farm value only	Improved acres and farm value only, with "tenant"[b]	Improved acres and unimproved acres only	Improved acres, unimproved acres, and farm value with "tenant"[b]	Schedule 1 "tenant"[c] also on Schedule IV	Schedule 1 "tenants"[c] not on Schedule IV
Pike	2	4 (2)						42	18		23		
Polk	4												
Pulaski	4												
Putnam	1	125 (77)											
Quitman	1	89 (76)											
Rabun	2	105 (101)											
Randolph	3												
Richmond	1	72 (26)		7 (5)		5 (4)	1						
Schley	1	28	2										
Screven	1							29					
Spalding	1					80 (73)					5		
Stewart	1	79 (69)		8							9		

194

Sumter	3		
Talbot	1	89 (87)	
Taliaferro	4		
Tattnall	1	46 (45)	
Taylor	3		
Telfair	5		
Terrell	4		
Thomas	3		
Towns	1	183 (174)	
Troup	2	64 (59)	71 (70)
Twiggs	1	182 (155)	
Union	2	194 (178)	
Upson	1	182 (140)	
Walker	1	419 (367)	
Walton	2	49 74	9 X
Ware	1	25 (24)	
Warren	1		53

(*continued*)

Appendix A (Continued)

County	Class	Standard convention[a]	Standard convention with tenant[b]	Improved acres only	Improved acres only, with "tenant"[b]	Farm value only	Farm value only, with "tenant"[b]	Improved acres and farm value only	Improved acres and farm value only, with "tenant"[b]	Improved acres and unimproved acres only	Improved acres, unimproved acres, and farm value with "tenant"[b]	Schedule I "tenant",[c] also on Schedule IV	Schedule I "tenants",[c] not on Schedule IV
Washington	1	151 (147)											
Wayne	1	3	4		5						2		X
Webster	1	60 (57)											
White	1	162 (158)											
Whitfield	5												
Wilcox	1	44 (41)											
Wilkes	2	2		10						1			
Wilkinson	1	208 (207)											
Worth	1										60		

Appendix B

TENANCY RATE (%) AND TYPE 3 AND TYPE 5 ESTIMATES BY COUNTY (SHORT METHOD), 1860

County	Class	Tenants[a] (%)	Type 3[b]	Type 5[b]
Appling	3			
Baker	2	27.7−	Low	High
Baldwin	2	21.0	Medium	Low
Banks	2	36.5		Low
Berrien	4			Low
Bibb	3			
Brooks	3			
Bryan	1	15.5	Low	Low
Bulloch	4			Low
Burke	4		Low	Low
Butts	3			
Calhoun	3			
Camden	3			
Campbell	1	23.9		def.[c]
Carroll	3			
Cass	3			
Catoosa	4			High
Charlton	3			
Chatham	3			
Chattahoochee	2	16.8+	High	Low
Chattooga	1	15.5		Low
Cherokee	2	22.9		Medium
Clarke	2	18.4+	Medium	Low
Clay	1	19.2	Low	Medium
Clayton	5			Low
Clinch	1	26.3		High
Cobb	2	2.8	Low	High

(*continued*)

Note: An explanation of the short-method estimating procedure and of types 3 and 5 can be found in chapter 2.

[a] Excludes types b and c.

[b] Low: 5 percent or less; medium: more than 5 percent but not more than 10 percent; high: more than 10 percent.

[c] The occupations of large numbers of heads of household are not entered in the enumerators' returns.

198 Appendix B

Appendix B (Continued)

County	Class	Tenants[a] (%)	Type 3[b]	Type 5[b]
Coffee	1	29.4	Low	Low
Colquitt	1	5.6		Low
Columbia	1	7.4		High
Coweta	4			Low
Crawford	2	16.4+	Medium	Low
Dade	3			
Dawson	1	30.2		Low
Decatur	1	9.4		Low
De Kalb	2	11.1+	High	High
Dooly	1	19.0	Low	Low
Dougherty	3			
Early	3			
Echols	1	31.4	Low	Low
Effingham	5			High
Elbert	1	16.9		High
Emanuel	2	14.3+	Medium	Low
Fannin	1	4.0		Low
Fayette	1	30.4		Low
Floyd	1	27.2		Low
Forsyth	3			
Franklin	2	5.0+	Medium	High
Fulton	1	11.0	Low	High
Gilmer	2	12.3+	Medium	High
Glascock	2	17.3+	Medium	Low
Glynn	3			
Gordon	4			High
Greene	1	8.5	Low	Medium
Gwinnett	1	22.1	Low	Medium
Habersham	1	37.8	Low	Low
Hall	4			High
Hancock	3			
Haralson	1	27.2		Medium
Harris	1	20.4	Low	Medium
Hart	1	19.1		Low
Heard	4			High

Appendix B (Continued)

County	Class	Tenants[a] (%)	Type 3[b]	Type 5[b]
Henry	3			
Houston	3			
Irwin	1	7.6		High
Jackson	1	5.0		Low
Jasper	2	17.6	Low	Low
Jefferson	2	11.4+		Low
Johnson	1	16.4		Low
Jones	3			
Laurens	1	19.1		Low
Lee	1	10.4		Medium
Liberty	3			
Lincoln	3			
Lowndes	1	15.3		Low
Lumpkin	1	35.0	Low	Low
McIntosh	3			
Macon	3			
Madison	1	26.4		Low
Marion	1	18.1		Low
Meriwether	1	23.1		Low
Miller	3			
Milton	1	34.9		Low
Mitchell	4			Low
Monroe	5			Low
Montgomery	1	4.2		Medium
Morgan	3			
Murray	4			
Muscogee	2	7.7+	Medium	High
Newton	1	24.3	Low	Low
Oglethorpe	1	10.7		Medium
Paulding	2	5.2+	Low	Medium
Pickens	1	36.3		Medium
Pierce	3			
Pike	2	13.0+	Medium	Low
Polk	4			High
Pulaski	4			Medium

(continued)

200 Appendix B

Appendix B (Continued)

County	Class	Tenants[a] (%)	Type 3[b]	Type 5[b]
Putnam	1	21.6	Low	Low
Quitman	1	33.9	Low	Low
Rabun	2	27.1		
Randolph	3			
Richmond	1	8.3	Low	Medium
Schley	1	14.4		Low
Screven	1	6.3		Low
Spalding	1	17.2		High
Stewart	1	11.6		Low
Sumter	3			
Talbot	1	15.6		High
Taliaferro	4			
Tattnall	1	9.7		Low
Taylor	3			
Telfair	5			Low
Terrell	4			
Thomas	3			
Towns	1	43.2	Low	Low
Troup	2	18.4+	Medium	Low
Twiggs	1	35.4	Low	Low
Union	2	28.9	Low	Low
Upson	1	22.9		Low
Walker	1	35.2		Low
Walton	2	15.5+		Low
Ware	1	11.9		Medium
Warren	1	11.6		Low
Washington	1	17.2		Low
Wayne	1	7.5		High
Webster	1	17.3		Low
White	1	37.5		Low
Whitfield	5			Low
Wilcox	1	17.0	Low	Low
Wilkes	2	3.4+	Medium	Low
Wilkinson	1	27.8		Medium
Worth	1	21.0	Low	Low

Appendix C

LONG-METHOD ESTIMATED TENANCY RATES (%) AT FOUR LEVELS BY MILITIA DISTRICT, SIXTEEN SAMPLE COUNTIES, 1860

County	Militia District	Level I	Level II	Level III	Level IV	Type 5 or Farm Laborers or Both as Percentage of Total Tenancy Candidates
Baker		17.0	20.3	29.2	33.3[a]	52.7
	Milford	23.7	27.6	34.1	34.9	30.0
	Newton	13.2	16.2	26.7	32.6	63.9
Baldwin		18.1	25.5	25.1	26.2[a]	13.2
	105	13.5	16.2	20.5	22.5	33.3
	115	27.0	30.2	30.9	32.8	17.4
	318	13.6	20.4	19.6	19.6	0.0
	319	16.3	20.4	21.2	21.2	9.1
	320	0.0	8.0	6.4	6.4	0.0
	321	26.8	37.5	33.8	33.8	0.0
	322	13.9	33.3	31.6	35.0	14.3
Clay		18.1	19.6	23.9	26.3[a]	38.9
	431	20.8	20.8	21.5	22.7	26.7
	749	16.9	19.1	23.1	26.4	39.5
	969	18.5	19.6	26.4	28.3	43.8
Coffee		27.3	28.5	29.6[a]	n.a.	6.8
	437	23.7	25.4	25.4	n.a.	6.2
	748	31.8	31.8	31.8	n.a.	0.0
	1026	26.3	30.3	30.3	n.a.	0.0
	1030	16.7	16.7	20.4	n.a.	22.2
	1127	25.0	25.0	29.2	n.a.	21.4
	1170	38.0	38.0	39.2	n.a.	5.0
Echols		20.7[b]	23.1[b]	23.1[b]	46.2[a,b]	65.0
	719	33.3	66.7	66.7	75.0	33.3
	904	25.0[b]	28.6[b]	28.6[b]	63.0[b]	76.5
	1058	19.4	19.4	19.4	37.0	58.8
	1211	18.5	20.4	20.4	37.7	57.7
Emanuel		13.2	14.4	17.3[a]	n.a.	22.2
	49	15.6	15.6	17.0	n.a.	12.5
	50	16.1	16.1	18.8	n.a.	16.7
	53	12.0	15.7	17.4	n.a.	18.8
	57	16.9	16.9	23.4	n.a.	33.3
	58	10.3	10.3	15.6	n.a.	41.7
	59	13.9	13.9	13.7	n.a.	0.0
	395	8.9	8.9	12.0	n.a.	30.0
	1208	14.8	20.4	24.6	n.a.	21.4

(*continued*)

Note: An explanation of the derivation of tenancy levels can be found in chapter 6.
[a] Maximum tenancy estimate for county.
[b] Type 7a records are included in the denominator.

Appendix C (Continued)

County	Militia District	Level I	Level II	Level III	Level IV	Type 5 or Farm Laborers or Both as Percentage of Total Tenancy Candidates
Gwinnett		20.1	21.0	27.8	32.6[a]	50.6
	316	19.8	19.8	26.4	30.1	49.0
	404	20.3	20.3	26.9	37.4	61.8
	405	0.0	0.0	21.2	29.7	100.0
	406	21.4	28.6	37.5	37.5	33.3
	407	21.7	21.7	29.4	36.3	56.1
	408	22.7	25.4	29.4	29.9	31.7
	444	16.5	17.5	24.0	28.7	54.0
	478	21.0	21.0	29.5	30.6	47.4
	544	13.5	13.5	22.0	34.3	70.8
	550	26.2	26.2	32.6	38.8	44.7
	562	21.1	22.6	25.0	32.7	43.1
	571	20.4	21.6	31.9	31.9	48.6
Harris		19.1	20.5	26.3[a]	n.a.	30.2
	672	12.8	14.3	17.6	n.a.	23.1
	679	13.6	16.7	27.2	n.a.	50.0
	695	10.4	10.4	15.8	n.a.	41.7
	696	17.0	17.0	22.0	n.a.	30.8
	703	16.2	16.2	31.3	n.a.	61.3
	707	22.2	25.0	27.6	n.a.	14.3
	717	25.8	27.0	31.2	n.a.	20.0
	781	5.3	7.0	23.2	n.a.	85.7
	782	25.5	29.4	29.4	n.a.	0.0
	786	25.7	27.0	30.4	n.a.	16.7
	920	9.4	10.9	13.6	n.a.	22.2
	934	26.5	26.5	26.5	n.a.	0.0
	1186	36.1	37.5	40.2	n.a.	18.2
Twiggs		31.3	31.5	34.9[a]	n.a.	14.4
	323	30.0	30.0	30.0	n.a.	0.0
	324	5.0	5.0	13.6	n.a.	66.7
	325	27.3	27.3	35.1	n.a.	30.8
	326	29.7	29.7	29.7	n.a.	0.0
	354	31.1	33.3	33.3	n.a.	0.0
	355	40.0	40.0	42.0	n.a.	8.0
	356	36.6	36.6	44.7	n.a.	28.6
	372	31.1	31.1	36.7	n.a.	22.2
	396	38.9	38.9	42.1	n.a.	12.5
	425	4.8	4.8	13.0	n.a.	66.7

Appendix D

OWSLEY SCHOOL COMPUTATIONS OF LANDOWNERS AS PERCENTAGE OF TOTAL HEADS OF HOUSEHOLD ENGAGED IN AGRICULTURE, BY ECONOMIC REGION, 1850 AND 1860

Economic Region	Landowners, 1850 (%)	Landowners, 1860 (%)
Alabama Pine Belt	44.16	72.60
Alabama Black Belt	76.98	80.56
Alabama Uplands	70.96	75.62
Georgia Pine Belt	—	81.07
Georgia Black Belt	74.13	76.44
Georgia Uplands	68.54	62.52
Louisiana Pine Belt and Prairies	61	73
Louisiana Sugar Bowl	80+	—
Louisiana Northeastern Alluvial	—	—
Louisiana Oak Uplands	68	—
Mississippi Pine Barrens	77.78	77.54
	(71.67)[a]	(71.24)[a]
Mississippi Delta-Loess	89.29	85.88
	(83.15)[a]	(75.33)[a]
Mississippi Northeastern Hills	83.46	74.32
	(77.99)[a]	(66.74)[a]
Tennessee Mountains and Valleys	59.67	60.25
Tennessee Highland Rim	68.50	67.79
Tennessee Central Basin	61.90	74.56
Tennessee Western Plateau	68.85	69.10
Tennessee Mississippi Bottom	67.25	78.43

Sources: Alabama and Georgia: Frank Lawrence Owsley, *Plain Folk of the Old South* (Chicago, 1965 [first published, 1949]), 155, 158, 171, 177, 182, and 189. Louisiana: ibid., 168; and Harry L. Coles, Jr., "Some Notes on Slave and Land Ownership in Louisiana, 1850–1860," *The Journal of Southern History* 9 (Aug. 1943): 386. Mississippi: Herbert Weaver, *Mississippi Farmers, 1850–1860* (Gloucester, Mass., 1968 [first published, 1945]), 36 and 64–66. Tennessee: Blanche Henry Clark, *The Tennessee Yeoman, 1840–1860* (New York, 1971 [first published, 1942]), 28.

Note: With the exception of the Louisiana Sugar Bowl and Mississippi, for which regional percentages are given in the source, the figures for this table are our own computations from county percentages weighted by the number of agricultural household heads in each of the Owsley School's sample counties. The figures do not, therefore, purport to estimate the true percentages for each region defined by the Owsley School authors. In the case of one county in the Alabama Black Belt and two counties in the Georgia Pine Belt, Owsley used county tax returns, one in Georgia being for 1862. Also in the Georgia Pine Belt for 1850, census returns for two of the four counties were unavailable to Owsley and for the remaining two counties were computed by him from tax returns for 1844 and 1853. This region is therefore excluded from the table for 1850. In the case of the Georgia Black Belt, two counties are excluded for 1860, since the census manuscript was not available to Owsley. We have excluded one county in the Alabama Black Belt for 1850, because Owsley's results, based on tax returns, were given for 1837, 1844, and 1856. Landholding percentages for Louisiana, with the exception of the Pine Belt and Prairies, were presented unsystematically by Harry L. Coles, Jr. Of the two parishes in the Northeastern Alluvial, Tensas had a landownership rate of 82 percent in 1850, while Catahoula had rates of 66 and 92 percent in 1850 and 1860 respectively. Since the number of farming household heads is not given, we cannot compute a weighted percentage for 1850. According to Coles, "more than 80 percent" of the farmers in the Sugar Bowl owned their own land in 1850. He noted further that the rate of landownership increased in all parishes except Washington in the Pine Belt, where it decreased, and West Feliciana in the Sugar Bowl, where it remained about the same. In the sources, parish percentages in Louisiana are rounded to the nearest whole number. The method of the Owsley School is discussed in greater detail in Chapter 2.

[a] The figures for Mississippi not in parentheses are Herbert Weaver's estimates and are too high. The figures in parentheses are our own revisions based on Weaver's data. See chapter 2.

Appendix E

FARM ACRES AND TOTAL SURFACE AREA, BY COUNTY, 1860

County	Farm Acres, 1860[a]	Total Surface Area, 1960 (Acres)[b]	Farm Acres as Percentage of Surface Area
Baker	162,605	227,840	71.4
Baldwin	159,826	169,600	94.2
Banks	128,475	147,840	86.9
Bibb	147,330	162,560	90.6
Brooks	266,936	318,080	83.9
Bryan	172,410	291,200	59.2
Butts	122,105	120,320	101.5
Calhoun	138,434	184,960	74.8
Camden	182,000	444,800	40.9
Cass	231,755	304,640	79.4
Catoosa	78,721	106,880	73.6
Chatham	147,136	321,280	45.8
Chattahoochee	127,931	161,920	79.0
Cherokee	179,653	273,920	65.6
Clay	117,893	143,360	82.2
Clayton	87,511	95,360	91.8
Cobb	176,617	222,720	79.3
Colquitt	75,749	360,320	21.0
Coweta	292,395	283,520	103.1
Crawford	180,660	201,600	89.6
Dade	55,783	107,520	51.9

Sources: U.S. Census Office, *Agriculture of the United States in 1860; Compiled from the Original Returns of the Eighth Census* (Washington, 1864), 22, 26; and U.S. Bureau of the Census, "Areas of Georgia: 1960," *Area Measurement Reports* (Washington, 1965).

Notes: Total farm area and surface area have been combined in four groups of counties that experienced either new county creation or significant boundary adjustment. Decatur, Grady, Seminole, and Thomas have a combined farm area of 644,342 acres and a combined surface area of 1,216,000 acres. Chattooga and Floyd have a combined farm area of 426,518 acres and a combined surface area of 531,840 acres. Lamar, Monroe, and Pike have a combined farm area of 509,869 acres and a combined surface area of 518,400 acres. Columbia, McDuffie, and Warren have a combined farm area of 503,780 acres and a combined surface area of 547,200 acres.

This table includes only those counties that underwent no significant boundary changes between 1860 and 1960. No reliable area estimates exist for the period before the most recent creation of new counties in Georgia in 1920 and 1924.

[a] Includes both improved and unimproved acres.

[b] Includes both land area and water area, the latter because of the recent creation of large reservoirs in several counties. Area given in square miles in the source has been converted to acres.

Appendix E (Continued)

County	Farm Acres, 1860[a]	Total Surface Area, 1960 (Acres)[b]	Farm Acres as Percentage of Surface Area
De Kalb	144,523	172,160	83.9
Dougherty	190,475	210,560	90.5
Early	189,383	343,040	55.2
Echols	55,884	272,000	20.5
Effingham	234,609	307,200	76.4
Fannin	84,050	256,000	32.8
Forsyth	127,960	155,520	82.3
Glascock	71,897	91,520	78.6
Glynn	108,317	297,600	36.4
Gordon	175,878	229,120	76.8
Greene	281,877	257,920	109.3
Hancock	327,667	310,400	105.6
Harris	296,089	302,720	97.8
Heard	178,063	193,280	92.1
Jasper	237,127	240,000	98.8
Jefferson	353,131	340,480	103.7
Jones	269,906	257,280	104.9
Laurens	302,584	519,040	58.3
Lee	199,012	229,120	86.8
Lincoln	141,158	163,200	86.5
Marion	151,898	233,600	65.0
Meriwether	307,088	319,360	96.2
Miller	59,827	183,680	32.6
Mitchell	110,222	327,040	33.7
Morgan	213,539	228,480	93.5
Murray	128,023	218,880	58.5
Muscogee	144,001	142,080	101.4
Oglethorpe	264,813	278,400	95.1
Paulding	92,548	203,520	45.5
Pickens	90,388	144,000	62.8
Polk	118,660	199,680	59.4
Putnam	225,276	224,000	100.6
Quitman	79,484	109,440	72.6
Rabun	139,472	240,000	58.1

(continued)

Appendix E (Continued)

County	Farm Acres, 1860[a]	Total Surface Area, 1960 (Acres)[b]	Farm Acres as Percentage of Surface Area
Randolph	212,214	279,040	76.0
Richmond	210,585	208,640	100.9
Schley	103,118	103,680	99.4
Spalding	112,245	128,640	87.2
Stewart	282,887	296,320	95.5
Sumter	263,069	314,880	83.5
Talbot	241,845	249,600	96.9
Taliaferro	104,707	124,800	83.9
Taylor	167,483	257,920	64.9
Terrell	148,564	210,560	70.6
Towns	62,908	110,080	57.1
Troup	259,771	286,080	90.8
Twiggs	232,409	233,600	99.5
Union	121,643	204,160	59.6
Upson	195,092	213,760	91.3
Walker	190,538	284,800	66.9
Washington	425,464	431,360	98.6
Webster	122,154	124,800	97.9
Whitfield	155,207	179,840	86.3
Wilkes	291,613	302,080	96.5
Wilkinson	249,079	293,120	85.0
Total	26,650,490	37,680,640	70.7

Appendix F

FREQUENCY DISTRIBUTION OF PERSONAL PROPERTY, SIXTEEN SAMPLE COUNTIES, 1860

Personal Property ($)	Class									
	1a		2a		3a		4		5	
	No.	%	No.	%	No.	%	No.	%	No.	%
					Baker County					
0	2	1.5	3	8.3	1	14.3	2	15.4	6	18.8
1–49			1	2.8					1	3.1
50–99										
100–499	18	13.4	15	41.7[a]	3	42.8[a]			9	28.1
500–999	14	10.4	7	19.4	1	14.3	1	7.7	3	9.4[b]
Cumulative		25.4		72.2		71.4		23.1		59.4
1,000–2,499	12	9.0	6	16.7			5	38.5[a]	3	9.4
2,500–4,999	20	14.9	2	5.6			3	23.1	4	12.5
5,000–9,999	18	13.4[a]	2	5.6	1	14.3	1	7.7	4	12.5
10,000–19,999	16	11.9							2	6.2
20,000–49,999	25	18.6			1	14.3	1	7.7		
50,000–99,999	9	6.7								
100,000+										
Total	134		36		7		13		32	
Mean	12,894.9		1,114.7		4,448.6		3,395.4		2,104.7	

[a]The interval containing the median.
[b]When the total cases is even and the median falls between intervals, this note identifies the higher interval.

(continued)

Appendix F (Continued)

Personal Property ($)	Class									
	1a		2a		3a		4		5	
	No.	%	No.	%	No.	%	No.	%	No.	%
					Baldwin County					
0					3	13.0				
1–49					1	4.3				
50–99	1	0.5	3	5.8	2	8.7				
100–499	22	10.6	6	11.5	7	30.4[a]	1	4.0		
500–999	17	8.2	29	55.8[a]	1	4.3	4	16.0	6	85.7[a]
Cumulative		19.3	3	5.8		60.9	1	4.0		85.7
1,000–2,499	20	9.7		78.8	4	17.4		24.0	1	14.3
2,500–4,999	20	9.7	4	7.7	3	13.0	1	4.0		
5,000–9,999	26	12.6[a]	1	1.9	1	4.3	2	8.0		
10,000–19,999	46	22.2	4	7.7	1	4.3	2	8.0		
20,000–49,999	36	17.4	2	5.8			1	4.0		
50,000–99,999	11	5.3					4	16.0[a]		
100,000+	8	3.9					8	32.0		
Total	207		52		23		1	4.0		
Mean	26,431.5		1,476.2		1,505.4		25		7	
							39,727.6		296.4	

208

	Class									
	1a		2a		3a		4		5	
Personal Property ($)	No.	%	No.	%	No.	%	No.	%	No.	%
					Clay County					
0			6	11.8			4	15.4	3	12.5
1–49			1	2.0					1	4.2
50–99	1	0.4	2	3.9					4	16.7
100–499	56	25.4	28	54.9[a]	3	75.0[a]	10	38.5[a]	12	50.0[a]
500–999	31	14.1	8	15.7			4	15.4	1	4.2
Cumulative		40.0		88.2		75.0		69.2		87.5
1,000–2,499	24	10.9[a]	2	3.9			2	7.7	1	4.2
2,500–4,999	28	12.7	3	5.9			1	3.8		
5,000–9,999	31	14.1	1	2.0	1	25.0				
10,000–19,999	23	10.4					1	3.8	1	4.2
20,000–49,999	22	10.0					1	3.8		
50,000–99,999	4	1.8					3	11.5		
100,000+										
Total	220		51		4		26		24	
Mean	7,290.8		584.5		1,356.2		11,906.8		1,072.8	

(continued)

Appendix F (Continued)

Personal Property ($)	Class									
	1a		2a		3a		4		5	
	No.	%	No.	%	No.	%	No.	%	No.	%
	Coffee County									
0	1	0.4			1	25.0				
1–49	2	0.8	7	7.6					3	42.8
50–99	4	1.6	17	18.5					1	14.3[a]
100–499	57	23.6	48	52.2[a]	2	50.0[a]	1	25.0	2	28.6
500–999	62	25.7[a]	17	18.5			1	25.0	1	14.3
Cumulative		52.3		96.7		75.0		50.0		100.0
1,000–2,499	46	19.1	1	1.1	1	25.0	1	25.0[b]		
2,500–4,999	30	12.4	1	1.1						
5,000–9,999	27	11.2	1	1.1						
10,000–19,999	6	2.5					1	25.0		
20,000–49,999	5	2.1								
50,000–99,999	1	0.4								
100,000+										
Total	241		92		4		4		7	
Mean	3,015.6		388.2		576.8		3,663.8		185.7	

210

Personal Property ($)	Class											
	1a		2a		3a		4		5		7a	
	No.	%	No.	%	No.	%	No.	%	No.	%	No.	%
	Echols County											
0					3	100.0[a]						
1–49												
50–99												
100–499	10	11.1	12	48.0							2	66.7[a]
500–999	20	22.2	8	32.0[a]							1	33.3
Cumulative		33.3		80.0		100.0						100.0
1,000–2,499	22	24.4[a]	2	8.0								
2,500–4,999	17	18.9	3	12.0								
5,000–9,999	12	13.3										
10,000–19,999	5	5.6										
20,000–49,999	4	4.4										
50,000–99,999												
100,000+												
Total	90		25		3						3	
Mean	3,730.9		1,013.8		0.0						448.0	

(*continued*)

Appendix F (Continued)

Personal Property ($)	Class									
	1a		2a		3a		4		5	
	No.	%	No.	%	No.	%	No.	%	No.	%
	Emanuel County									
0	3	0.6	4	5.6	1	16.7			5	22.7
1–49	1	0.2	3	4.2			1	7.1	2	9.1
50–99	7	1.5	6	8.4					3	13.6
100–499	125	27.2	41	57.7[a]	3	50.0[a]	10	71.4[a]	11	50.0[a]
500–999	87	19.0	5	7.0			2	14.3		
Cumulative		48.6		83.1		66.7		92.8		95.4
1,000–2,499	93	20.3[a]	6	8.4	1	16.7			1	4.5
2,500–4,999	64	13.9	4	5.6	1	16.7	1	7.1		
5,000–9,999	45	9.8	2	2.8						
10,000–19,999	26	5.7								
20,000–49,999	7	1.5								
50,000–99,999	1	0.2								
100,000+										
Total	459		71		6		14		22	
Mean	2,939.8		702.0		750.0		452.1		144.4	

212

	Class									
	1a		2a		3a		4		5	
Personal Property ($)	No.	%	No.	%	No.	%	No.	%	No.	%
					Greene County					
0	20	5.5	1	3.2	2	33.3	5	20.0	7	16.3
1–49			1	3.2						
50–99	1	0.3	1	3.2					1	2.3
100–499	10	2.8	16	51.6[a]	1	16.7	2	8.0	22	51.2[a]
500–999	10	2.8	2	6.4	1	16.7[b]				
Cumulative		11.3		67.7		66.7		28.0		69.8
1,000–2,499	26	7.2	2	6.4			3	12.0	3	7.0
2,500–4,999	44	12.2	3	9.7			4	16.0[a]	4	9.3
5,000–9,999	71	19.6[a]	3	9.7			7	28.0	4	9.3
10,000–19,999	89	24.6	2	6.4	1	16.7	3	12.0	1	2.3
20,000–49,999	77	21.3			1	16.7	1	4.0	1	2.3
50,000–99,999	13	3.6								
100,000+	1	0.3								
Total	362		31		6		25		43	
Mean	14,156.8		2,253.7		7,100.0		5,450.8		1,977.3	

(*continued*)

Appendix F (Continued)

Personal Property ($)	Class									
	1a		2a		3a		4		5	
	No.	%	No.	%	No.	%	No.	%	No.	%
					Gwinnett County					
0	3	0.4	10	5.2			2	2.3	18	14.6
1–49	5	0.7	9	4.7			7	8.1	19	15.4
50–99	13	1.7	26	13.5	1	12.5	8	9.3	22	17.9
100–499	323	42.7	126	65.3[a]			54	62.8[a]	58	47.2[a]
500–999	134	17.7[a]	12	6.2	1	12.5	8	9.3	6	4.9
Cumulative		63.1		94.8	6	75.0[a]		91.9		100.0
1,000–2,499	93	12.3	7	3.6		100.0	3	3.5		
2,500–4,999	68	9.0	2	1.0			3	3.5		
5,000–9,999	65	8.6	1	0.5			1	1.2		
10,000–19,999	32	4.2								
20,000–49,999	18	2.4								
50,000–99,999	2	0.3								
100,000+	1	0.1								
Total	757		193		8		86		123	
Mean	2,778.6		297.7		183.8		427.2		127.7	

	Class									
	1a		2a		3a		4		5	
Personal Property ($)	No.	%	No.	%	No.	%	No.	%	No.	%
					Habersham County					
0	7	1.4	4	1.4	3	75.0[a]			3	6.2
1–49			11	3.8			1	33.3	20	41.7
50–99	14	2.9	48	16.6			1	33.3[a]	16	33.3[a]
100–499	227	47.0[a]	208	72.0[a]	1	25.0	1	33.3	9	18.8
500–999	97	20.1	11	3.8						
Cumulative		71.4		97.6		100.0		100.0		100.0
1,000–2,499	70	14.5	6	2.1						
2,500–4,999	27	5.6	1	0.3						
5,000–9,999	18	3.7								
10,000–19,999	13	2.7								
20,000–49,999	9	1.9								
50,000–99,999										
100,000+	1	0.2								
Total	483		289		4		3		48	
Mean	1,976.2		216.1		50.0		58.3		48.2	

(*continued*)

Appendix F (Continued)

Personal Property ($)	Class									
	1a		2a		3a		4		5	
	No.	%	No.	%	No.	%	No.	%	No.	%
					Harris County					
0			8	5.0	10	83.3[a]	1	4.2	4	5.3
1–49	1	0.2	2	1.2					5	6.7
50–99	2	0.3	1	0.6					8	10.7
100–499	78	12.0	71	44.1[a]			5	20.8	39	52.0[a]
500–999	88	13.5	25	15.5			9	37.5[a]	8	10.7
Cumulative		25.9		66.4		83.3		62.5		85.3
1,000–2,499	76	11.6	17	10.6	1	8.3	1	4.2	1	1.3
2,500–4,999	88	13.5[a]	20	12.4	1	8.3	1	4.2	2	2.7
5,000–9,999	101	15.5	7	4.3			2	8.3	6	8.0
10,000–19,999	120	18.4	9	5.6			1	4.2	2	2.7
20,000–49,999	75	11.5	1	0.6			4	16.7		
50,000–99,999	21	3.2								
100,000+	2	0.3								
Total	652		161		12		24		75	
Mean	10,228.0		2,091.5		462.1		6,379.2		1,008.1	

216

Personal Property ($)	1a		2a		3a		4		5	
	No.	%	No.	%	No.	%	No.	%	No.	%
					Houston County					
0	6	1.2	2	3.3			1	8.3	18	30.5
1–49	2	0.4							2	3.4
50–99	1	0.2	1	1.7			2	16.7	2	3.4
100–499	59	11.7	11	18.3			7	58.3[a]	26	44.1[a]
500–999	27	5.4	5	8.3					3	5.1
Cumulative		18.9		31.7				83.3		86.4
1,000–2,499	47	9.3	9	15.0					5	8.5
2,500–4,999	54	10.7	11	18.3[a]			1	8.3	1	1.7
5,000–9,999	83	16.5[a]	12	20.0					2	3.4
10,000–19,999	85	16.9	4	6.7						
20,000–49,999	106	21.1	5	8.3			1	8.3		
50,000–99,999	27	5.4								
100,000+	6	1.2								
Total	503		60				12		59	
Mean	16,163.8		5,497.1				2,807.8		524.5	

(*continued*)

Appendix F (Continued)

Personal Property ($)	Class									
	1a		2a		3a		4		5	
	No.	%	No.	%	No.	%	No.	%	No.	%
					Thomas County					
0	3	1.0	7	22.6	6	40.0	4	5.2	78	52.7[a]
1–49							1	1.3	1	0.7
50–99							2	2.6	1	0.7
100–499	32	11.2	13	41.9[a]	2	13.3[a]	22	28.6	45	30.4
500–999	24	8.4			1	6.7	10	13.0[a]	3	2.0
Cumulative		20.7		64.5		60.0		50.6		86.5
1,000–2,499	31	10.9	4	12.9	2	13.3	15	19.5	9	6.1
2,500–4,999	36	12.6	5	16.1			6	7.8	8	5.4
5,000–9,999	58	20.4[a]	1	3.2	2	13.3	5	6.5		
10,000–19,999	46	16.1	1	3.2	2	13.3	7	9.1	3	2.0
20,000–49,999	42	14.7					3	3.9		
50,000–99,999	11	3.8								
100,000+	2	0.7					2	2.6		
Total	285		31		15		77		148	
Mean	12,329.8		1,494.8		2,528.0		9,545.1		662.0	

	Class									
	1a		2a		3a		4		5	
Personal Property ($)	No.	%	No.	%	No.	%	No.	%	No.	%
				Towns County						
0	3	1.3	27	16.6	3	60.0[a]			3	50.0
1–49	14	6.1	28	17.2						
50–99	16	6.9	29	17.8[a]					1	16.7[b]
100–499	121	52.4[a]	67	41.1	2	40.0	1	100.0	2	33.3
500–999	37	16.0	9	5.5						
Cumulative		82.7		98.2		100.0		100.0		100.0
1,000–2,499	22	9.5	2	1.2						
2,500–4,999	10	4.3	1	0.6						
5,000–9,999	6	2.6								
10,000–19,999	1	0.4								
20,000–49,999	1	0.4								
50,000–99,999										
100,000+										
Total	231		163		5		1		6	
Mean	845.8		163.6		60.0				56.7	

(*continued*)

Appendix F (Continued)

Personal Property ($)	1a		2a		Class 3a		4		5	
	No.	%	No.	%	No.	%	No.	%	No.	%
					Twiggs County					
0			1	0.7					10	43.5
1–49			2	1.5					1	4.3
50–99	4	1.4	10	7.4					9	39.1a
100–499	51	17.3	70	51.5a	1	100.0			3	13.0
500–999	29	9.9	17	12.5						
Cumulative		28.6		73.5		100.0				100.0
1,000–2,499	35	11.9	11	8.1						
2,500–4,999	28	9.5a	10	7.4						
5,000–9,999	39	13.3	13	9.6						
10,000–19,999	47	16.0	2	1.5						
20,000–49,999	39	13.3								
50,000–99,999	16	5.4								
100,000+	6	2.0								
Total	294		136		1				23	
Mean	14,923.8		1,405.9						41.1	

	Class									
	1a		2a		3a		4		5	
Personal Property ($)	No.	%	No.	%	No.	%	No.	%	No.	%
					Wilkes County					
0	7	2.1	7	53.8[a]	8	28.6	2	8.7	8	42.1
1–49										
50–99							1	4.3		
100–499	33	9.9	2	15.4	4	14.3	4	17.4		
500–999	13	3.9	1	7.7	3	10.7[a]	1	4.3		
Cumulative		15.9		76.9		53.6		34.8		42.1
1,000–2,499	26	7.8			4	14.3			3	15.8[a]
2,500–4,999	32	9.6	1	7.7	3	10.7	3	13.0	3	15.8
5,000–9,999	48	14.4	2	15.4	3	10.7	2	8.7[a]	2	10.5
10,000–19,999	67	20.1[a]			3	10.7	1	4.3	3	15.8
20,000–49,999	75	22.5					4	17.4		
50,000–99,999	23	6.9					4	17.4		
100,000+	9	2.7					1	4.3		
Total	333		13		28		23		19	
Mean	20,785.4		1,103.8		3,004.5		24,912.6		3,213.2	

(continued)

Appendix F (Continued)

Worth County

Personal Property ($)	1a No.	1a %	2a No.	2a %	3a No.	3a %	4 No.	4 %	5 No.	5 %
0			2	3.8	1	9.1				
1–49	1	0.5	1	1.9					1	16.7
50–99			4	7.7			1	14.3		
100–499	55	25.9	33	63.5[a]	7	63.6[a]	4	57.1[a]	3	50.0[a]
500–999	63	29.7[a]	5	9.6					1	16.7
Cumulative		56.1		86.5		72.7		71.4		100.0
1,000–2,499	46	21.7	1	1.9	2	18.2				
2,500–4,999	16	7.5	4	7.7	1	9.1	1	14.3		
5,000–9,999	18	8.5	1	1.9			1	14.3		
10,000–19,999	9	4.2								
20,000–49,999	4	1.9	1	1.9						
50,000–99,999										
100,000+										
Total	212		52		11		7		6	
Mean	2,582.1		1,045.4		637.3		1,411.4		276.0	

Appendix G

Appendix G

STATISTICS BY MILITIA DISTRICT, TYPES 1A, 2A, 3A, and 6A, SIXTEEN SAMPLE COUNTIES, 1860

County	Militia District	Number of Cases				Total Operators (%)				Improved Acres (%)				Mean Improved Acres				
		1a	3a	6a	2a	1a	3a	6a	2a	1a	3a	6a	2a	1a	3a	6a	2a	All Operators
Baker	Milford	134	7	35	36	63.2	3.3	16.5	17.0	57.4	0.6	39.6	2.4	247.2	48.6	662.4	43.4	277.5
		42	3	13	18	55.3	3.9	17.1	23.7	52.2	0.6	43.1	4.1	249.8	41.7	687.5	52.5	273.6
	Newton	92	4	22	18	67.6	2.9	16.2	13.2	60.0	0.6	37.8	1.6	246.1	53.8	648.6	35.4	279.5
Baldwin	105	210	23	21	56	67.7	7.4	6.8	18.1	75.4	1.7	18.3	4.6	138.8	28.8	336.4	31.5	124.7
	115	30	1	2	5	81.1	2.7	2.7	13.5	94.2	0.4	1.7	3.7	147.0	20	80	34.6	126.6
	318	42	2	2	17	66.7	3.2	3.2	27.0	79.1	0.4	12.7	7.8	115.8	12.5	390.0	28.0	97.6
	319	26	3	9	6	59.1	6.8	20.4	13.6	55.4	2.1	40.4	2.1	259.0	85.0	546.7	42.5	276.5
	320	35	2	4	8	71.4	4.1	8.2	16.3	83.7	0.6	9.4	6.3	183.1	23.0	180.0	60.2	156.3
	321	23	2	0	0	92.0	8.0			99.5	0.5			92.3	5.0			85.3
	322	31	6	4	15	55.4	10.7	7.1	26.8	79.4	3.2	9.3	8.1	84.3	17.8	76.2	17.8	58.8
		23	7	1	5	63.9	19.4	2.8	13.9	77.9	7.7	10.1	4.3	87.1	28.4	260	22.0	71.4
Clay		220	4	6	51	78.3	1.4	2.1	18.1	93.6	0.1	6.3	n.d.	170.3	40	412.5	n.d.	143.5
	431	40	0	2	11	75.5		3.8	20.8	91.9		8.1	n.d.	179.1		315.0	n.d.	147.0
	749	108	3	2	23	79.4	2.2	1.5	16.9	93.6	0.2	6.2	n.d.	199.7	40	715.0	n.d.	171.9
	969	72	1	2	17	78.3	1.1	2.2	18.5	95.2	def.	4.8	n.d.	119.2	def.	207.5	n.d.	98.2
Coffee		241	4	0	92	71.5	1.2		27.3	89.0	1.7		9.4	46.5	52.0		14.5	38.7
	437	44	1	0	14	74.6	1.7		23.7	83.6	6.3		10.1	40.3	130		16.1	36.3
	748	45	0	0	21	68.2			31.8	87.9			12.1	38.7			11.8	30.3
	1026	53	3	0	20	69.7	3.9		26.3	88.5	2.9		8.6	46.0	26.0		14.6	38.1
	1030	35	0	0	7	83.3			16.7	94.0			6.0	38.1			16.4	35.3
	1127	33	0	0	11	75.0			25.0	94.5			5.5	81.9			17.4	68.1
	1170	31	0	0	19	62.0			38.0	83.1			16.9	38.9			14.4	30.2
Echols	719	90	3	0	25	76.3	2.5		21.2	86.5	2.4		11.1	68.3	56.7		31.7	60.3
	904	1	1	0	1	33.3	33.3		33.3	48.1	44.4		7.4	65	60		10	45.0
	1058	17	0	0	7	68.0			28.0	84.0			11.2	71.2	70		23.0	57.7
	1211	29	0	0	7	80.6			19.4	85.5			14.5	74.4			52.3	70.1
		43	1	0	10	79.6	1.8		18.5	90.2	1.3		8.5	63.1	40		25.6	55.8
Emanuel		459	6	0	71	85.6	1.1		13.2	99.5	0.5		n.d.	85.1	32.3		n.d.	73.1
	49	76	0	0	14	84.4			15.6	100.0			n.d.	118.9			n.d.	100.4
	50	26	0	0	5	83.9			16.1	100.0			n.d.	50.0			n.d.	42.0
	53	70	3	0	10	84.3	3.6		12.0	99.0	1.0		n.d.	87.1	19.7		n.d.	73.9
	57	49	0	0	10	83.0			16.9	100.0			n.d.	125.3			n.d.	103.3
	58	61	0	0	7	89.7			10.3	100.0			n.d.	67.5			n.d.	60.4
	59	62	0	0	10	86.1			13.9	100.0			n.d.	65.9			n.d.	56.8
	395	72	0	0	7	91.1			8.9	100.0			n.d.	65.5			n.d.	59.6
	1208	43	3	0	8	79.6	5.6		14.8	96.4	3.6		n.d.	84.1	45.0		n.d.	69.5
Greene		362	6	11	31	88.3	1.5	2.7	7.6	94.6	1.3	4.1	def.	316.3	260.0	442.9	def.	295.1

224

	316	101	0	25	80.2			19.8			def.	62.8	27.1	def.	62.5
	404	51	0	13	79.7			20.3			n.d.	53.3		n.d.	42.7
	407	65	0	18	78.3			21.7			n.d.	91.5		n.d.	72.0
	408	82	3	25	74.5	2.7		22.7			7.1	107.7	33.3	29.9	90.8
	444	80	1	16	82.5	1.0		16.5			def.	98.9	25	def.	74.5
	478	75	0	20	78.9			21.0		1.3	8.6	47.4		25.4	43.9
	544	45	0	7	86.5			13.5		0.6	def.	91.1		def.	72.2
	550	59	0	21	73.8			26.2			3.3	73.2		18.7	66.8
	562	99	2	27	77.3	1.6		21.0		0.4	n.d.	88.8	13.5	n.d.	64.9
	571	69	0	18	78.4	1.1		20.4		0.4	8.8	66.6	40	23.8	56.8
											n.d.	131.2		n.d.	103.0
Habersham	483	4	3	289	62.0	0.5	0.4	37.1	98.2	0.2	1.6	65.1	16.5	n.d.	40.9
Harris	652	12	18	161	77.3	1.4	2.1	19.1	95.4	1.9	2.7	231.1	248.3	n.d.	187.5
	660	1	0	9	85.7	1.4		12.8	96.6	3.3		244.9	500	n.d.	216.6
	679	54	2	9	81.8	3.0	1.5	13.6	97.1	2.2	0.8	239.7	145.0	n.d.	202.0
	695	59	0	7	88.0		1.5	10.4	99.8		0.2	297.3		n.d.	262.4
	696	42	0	9	79.2		3.8	17.0	99.2		0.8	300.6	100	n.d.	238.9
	703	60	2	12	81.1		2.7	16.2	94.8		5.1	162.9	37	n.d.	141.2
	707	54	2	16	75.0	2.8		22.2	98.2	1.8		155.6	50.0	n.d.	117.7
	717	62	1	23	69.7	1.1	3.4	25.8	96.3	0.8	2.8	217.2	530	n.d.	158.9
	781	51	2	3	89.5	1.8	3.5	5.3	83.4	4.0	12.5	182.6	197.5	n.d.	195.8
	782	29	7	13	56.9	3.9	13.7	25.5	79.0	2.5	18.5	253.0	700.0	n.d.	182.1
	786	54	1	19	73.0	1.4		25.7	93.0	7.0		244.4	115.0	n.d.	189.7
	920	57	0	6	89.1	1.6		9.4	99.2	0.8		213.3	245.7	n.d.	191.2
	934	25	0	9	73.5			26.5	100.0			576.6	950	n.d.	424.0
	1186	45	1	26	62.5	1.4		36.1	96.2	3.8		106.2	100	n.d.	69.0
													190		
Houston	503		0	3	60	88.9	0.5	10.6	98.9	5.5	1.1	384.2	666.7	n.d.	345.1
Thomas	285	15	5	31	84.8	4.5	1.5	9.2	87.2		6.6	231.2	270.7	962.0	def.	223.8
Towns	231	5	2	163	57.6	1.2	0.5	40.6	98.5	1.0	0.5	58.1	25.0	32.5	n.d.	33.3
Twiggs	294	1	4	136	67.6	0.2	0.9	31.3	97.8	0.1	2.0	361.4	n.a.	700.0	n.a.	245.9
	323	41	1	18	68.3		1.7	30.0	91.7		8.2	190.2		700	def.	141.7
	324	19	0	1	95.0			5.0	100.0			672.2			n.d.	638.6
	325	24	0	9	72.7			27.3	100.0			730.5			n.d.	511.3
	326	26	0	11	70.3			29.7	100.0			295.1			n.d.	204.9
	354	30	0	14	66.7	2.2		31.1	98.9	1.1		342.7			n.d.	219.8
	355	69	0	46	60.0			40.0	100.0			175.7	100		n.d.	104.2
	356	25	0	15	61.0		2.4	36.6	91.5		8.5	401.9		900	n.d.	263.6
	372	30	1	14	66.7		2.2	31.1	100.0		def.	321.4		def.	n.d.	222.0
	396	11	0	7	61.1			38.9	100.0			1,203.1			n.d.	641.7
	425	19	0	1	90.5		4.8	4.8	93.8		6.2	444.0		500	n.d.	423.6
Wilkes	333	28	8	13	87.2	7.3	2.1	3.4	93.8	3.3	2.2	366.5	160.3	356.2	86.0	343.5
Worth	212	11	11	52	74.1	3.8	3.8	18.2	72.7	1.9	13.3	77.5	39.2	273.6	52.3	79.0

(*continued*)

Appendix G (Continued)

County	Militia District	Unimproved Acres (%)			Mean Unimproved Acres					% Improved of Total Acres				
		1a	3a	6a	2a	1a	3a	6a	2a	All Operators	1a	3a	6a	2a	All Operators
Baker	Milford	65.5	0.3	34.2	n.d.	513.2	46.4	1,039.4	n.d.	494.8	def.	def.	def.	n.d.	35.5
	Newton	69.9	0.6	29.5	n.d.	567.2	70.0	797.5	n.d.	444.8	def.	def.	def.	n.d.	37.1
		63.5	0.2	36.3	n.d.	489.7	28.8	1,171.4	n.d.	521.6	def.	def.	def.	n.d.	34.7
Baldwin	105	82.5	2.2	15.3	n.d.	366.1	88.6	680.8	n.d.	300.7	def.	def.	def.	n.d.	29.3
	115	97.9	0.0	2.1	n.d.	316.5	n.a.	200	n.d.	262.0	def.	def.	def.	n.d.	32.6
	318	89.5	0.0	10.5	n.d.	287.7	0	710.0	n.d.	214.3	def.	def.	def.	n.d.	31.3
	319	59.2	1.6	39.2	n.d.	510.9	117.3	978.8	n.d.	510.1	def.	def.	def.	n.d.	35.1
	320	90.8	0.4	8.8	n.d.	456.1	35.0	388.8	n.d.	358.9	def.	def.	def.	n.d.	30.3
	321	100.0	0.0		n.d.	212.2	0.5		n.d.	187.7	def.	def.		n.d.	30.4
	322	85.4	2.0	12.6	n.d.	396.9	48.3	453.2	n.d.	257.3	def.	def.	def.	n.d.	18.6
		82.9	12.4	4.7	n.d.	385.9	189.3	500	n.d.	297.2	def.	def.	def.	n.d.	19.4
Clay	431	94.8	0.1	5.1	n.d.	341.9	90	675.8	n.d.	285.5	33.2	30.8	37.9	n.d.	33.5
	749	86.7		13.3	n.d.	383.8		1,175.0	n.d.	334.0	31.8		21.1	n.d.	30.6
	969	96.8	0.3	2.9	n.d.	300.2	90	485.0	n.d.	249.8	40.0	30.8	59.6	n.d.	40.8
		97.4	def.	2.6	n.d.	381.3	def.	367.5	n.d.	309.8	23.6		36.1	n.d.	24.0
Coffee	437	98.7	1.3		0.0	1,041.9	808.0		def.	751.3	def.	def.		def.	4.7
	748	95.6	4.4		n.d.	926.3	1,830		n.d.	718.3	def.	def.		n.d.	4.7
	1026	100.0			n.d.	1,270.0			n.d.	859.7	def.			n.d.	3.4
	1030	97.0	2.9		n.d.	889.5	467.3		n.d.	635.4	def.	def.		n.d.	5.4
	1127	100.0			n.d.	847.2			n.d.	702.5	def.			n.d.	4.6
	1170	100.0			n.d.	1,474.1			n.d.	1,105.6	def.			n.d.	5.6
		99.9			0.1	887.5			def.	860.6	def.			def.	5.0
Echols	719	97.0	3.0		n.d.	525.9	490.0		n.d.	413.6	def.	def.		n.d.	12.7
	904	100.0	0.0		n.d.	110	0		n.d.	36.7	def.	def.		n.d.	55.1
	1058	93.1	6.9		n.d.	421.2	530		n.d.	307.6	def.	def.		n.d.	15.8
	1211	100.0			n.d.	548.4			n.d.	441.8	def.			n.d.	13.7
		96.2	3.7		n.d.	561.8	940		n.d.	464.8	def.	def.		n.d.	10.7
Emanuel	49	99.8	0.1		n.d.	1,028.7	111.7		n.d.	880.8	7.6	22.4		n.d.	7.6
	50	100.0			n.d.	1,385.8			n.d.	1,170.2	7.9			n.d.	7.9
	53	100.0			n.d.	1,082.1			n.d.	907.6	4.4			n.d.	4.4
	57	99.6	0.4		n.d.	1,085.8	101.7		n.d.	917.4	7.4	16.2		n.d.	7.4
	58	100.0			n.d.	907.9			n.d.	748.6	12.1			n.d.	12.1
	59	100.0			n.d.	1,044.8			n.d.	935.6	6.1			n.d.	6.1
	395	100.0			n.d.	943.4			n.d.	812.4	6.5			n.d.	6.5
	1208	98.8	1.2		n.d.	909.1	121.7		n.d.	827.5	6.7	27.0		n.d.	6.7
					n.d.	703.3			n.d.	566.8	10.7			n.d.	10.9
Greene		96.5	1.2	2.2	n.d.	418.7	315.2	318.4	n.d.	382.6	def.	def.	def.	n.d.	43.6

	316	33.3	3.1			def.	211.0	21.8		177.2	def.		28.1
	404	100.0				n.d.	158.4			127.0	25.2		25.2
	407	100.0				n.d.	310.8			244.6	22.7		22.7
	408	99.8	0.2			n.d.	227.8			178.4	def.		33.7
	444	99.9	0.1			n.d.	175.2	10.0		130.1	def.	def.	36.3
	478	100.0				n.d.	128.7	10		106.0	def.	def.	28.8
	544	100.0				n.d.	234.0			184.7	def.		28.1
	550	100.0				def.	171.0			149.0	def.		30.5
	562	99.8	0.2			n.d.	369.9	19.0		270.3	19.4		19.4
	571	99.9	0.1			def.	195.5	20		154.1	def.	def.	26.9
						n.d.	247.4			194.2	34.4	66.7	34.4
Habersham		97.4	0.2	2.3	n.d.	n.d.	372.7	102.2	1,416.7	235.8	14.8	13.9	14.8
Harris	672	87.9	2.3	9.8	n.d.	n.d.	192.2	269.6	860.0	169.0	54.6	47.9	52.6
	679	95.4	4.6		n.d.	n.d.	247.2	700		221.6	49.8	41.7	49.4
	695	97.0	2.7	0.2	n.d.	n.d.	151.1	115.0	20	127.4	61.3	55.8	61.3
	696	100.0		0.0	n.d.	n.d.	146.3		n.a.	128.8	67.0		67.1
	703	97.8		2.2	n.d.	n.d.	213.3		100.0	172.0	58.5		58.1
	707	96.5		3.5	n.d.	n.d.	139.0		300	118.3	54.0		54.4
	717	98.2	1.8		n.d.	n.d.	200.0	97.5		151.4	43.7	43.5	43.7
	781	98.2	0.9	1.0	n.d.	n.d.	321.3	180	97.5	230.6	40.3	40.0	40.8
	782	42.1	2.0	55.9	n.d.	n.d.	166.6	400	5,650.0	354.4	52.3	52.9	35.6
	786	75.4	2.8	21.8	n.d.	n.d.	208.0	110.0	249.3	156.8	54.9	51.1	53.7
	920	88.5	11.5		n.d.	n.d.	177.8	1,200		145.1	57.9	44.2	56.7
	934	98.8	1.2		n.d.	n.d.	149.9	100		134.8	58.7	50.0	58.6
	1186	100.0			n.d.	n.d.	281.5			207.0	67.2		67.2
		99.8	0.2		n.d.	n.d.	132.8	10		83.2	44.4	95.0	45.3
Houston		98.6		1.4	n.d.	n.d.	381.7		871.3	344.3	50.1		50.1
Thomas		89.0	5.5	5.2	0.3	def.	484.7	554.1	1,566.0	460.1	32.2	32.8	32.7
Towns		96.4	1.4	2.2	n.d.	n.d.	211.3	141.0	527.5	123.9	21.6	15.1	21.2
Twiggs	323	97.1	0.1	2.9	n.d.	n.d.	450.5	n.a.	1,241.7	310.2	44.6	54.0	44.4
	324	91.3		8.7	n.d.	n.d.	223.5		875	167.3	46.0		45.8
	325	100.0			n.d.	n.d.	952.2			904.6	41.4		41.4
	326	100.0			n.d.	n.d.	604.0			428.6	56.5		56.5
	354	99.2	0.8		n.d.	n.d.	404.4			280.8	42.2		42.2
	355	100.0			n.d.	n.d.	396.8	85		253.7	46.3	54.0	46.4
	356	78.1		21.9	n.d.	n.d.	209.7			125.1	45.3		45.3
	372	100.0		def.	n.d.	n.d.	370.2		2,600	289.2	51.6		47.5
	396	100.0			n.d.	n.d.	752.4		def.	519.5	29.9		29.9
	425	97.9		2.0	n.d.	n.d.	988.0		250	526.9	54.9		54.9
					n.d.	n.d.	702.4			641.6	38.7		39.8
Wilkes		94.6	3.0	2.1	0.2	34.5	458.9	180.5	421.5	425.9	44.4	47.0	44.6
Worth		81.2	2.3	6.6	9.8	221.3	447.6	242.4	705.0	408.5	14.8	13.9	16.2

(*continued*)

Appendix G (*Continued*)

County	Militia District	Farm Value (%)			Mean Farm Value ($)					Farm Value/Acre				
		1a	3a	6a	2a	1a	3a	6a	2a	All Operators	1a	3a	6a	2a	All Operators
Baker	Milford	54.2	0.6	45.2	n.d.	7,057.1	1,385.7	21,295.6	n.d.	8,008.9	def.	def.	def.	n.d.	10.06
	Newton	59.9	0.5	39.6	n.d.	7,154.6	733.3	14,958.3	n.d.	6,388.4	def.	def.	def.	n.d.	8.81
		51.9	0.6	47.4	n.d.	7,013.5	1,875.0	24,752.3	n.d.	8,900.8	def.	def.	def.	n.d.	10.65
Baldwin	105	81.3	1.2	17.4	n.d.	3,898.9	903.6	8,361.9	n.d.	3,345.6	def.	def.	def.	n.d.	7.66
	115	98.5	def.	1.4	n.d.	3,153.3	def.	1,400	n.d.	2,666.7	def.	def.	def.	n.d.	6.69
	318	87.4	def.	12.6	n.d.	2,550.7	def.	7,750.0	n.d.	2,010.3	def.	def.	def.	n.d.	6.25
	319	59.9	0.8	39.3	n.d.	6,811.5	1,200.0	12,921.1	n.d.	6,878.8	def.	def.	def.	n.d.	8.58
	320	89.4	0.5	10.0	n.d.	5,006.3	1,000	4,927.5	n.d.	4,081.9	def.	def.	def.	n.d.	7.77
	321	99.0	1.0			3,596.7	425.0			3,343.0	def.	def.			11.92
	322	84.5	1.2	14.3	n.d.	3,468.9	500.0	4,550.0	n.d.	2,400.7	def.	def.	def.	n.d.	7.21
		86.7	8.0	5.2	n.d.	3,237.0	1,150.0	4,500	n.d.	2,452.8	def.	def.	def.	n.d.	6.50
Clay	431	92.9	0.1	7.0	n.d.	3,218.1	700	8,905.0	n.d.	2,741.4	6.34	5.38	8.18	n.d.	6.44
	749	90.0		10.0	n.d.	4,276.1		9,465.0	n.d.	3,584.4	7.60		6.35	n.d.	7.45
	969	93.2	0.2	6.6	n.d.	3,509.9	700	13,500.0	n.d.	3,035.6	7.02	5.38	11.25	n.d.	7.20
		95.5	def.	4.5	n.d.	2,192.6	def.	3,750	n.d.	1,817.2	4.46	def.	6.52	n.d.	4.53
Coffee	437	98.2	1.8		n.d.	1,147.3	1,250.0		n.d.	829.7	def.	def.		n.d.	1.05
	748	92.4	7.6		n.d.	996.4	3,500		n.d.	799.0	def.	def.		n.d.	1.06
	1026	100.0			n.d.	970.2			n.d.	651.8	def.			n.d.	0.72
	1030	97.6	2.3		n.d.	1,225.5	500.0		n.d.	864.9	def.	def.		n.d.	1.27
	1127	100.0			n.d.	1,266.5			n.d.	1,050.2	def.			n.d.	1.43
	1170	100.0			n.d.	1,565.2			n.d.	1,173.9	def.			n.d.	1.00
		100.0			n.d.	898.4			n.d.	557.0	def.			n.d.	0.96
Echols	719	95.6	1.8		2.5	2,300.0	1,300.0		220.0	1,833.9	def.	def.		n.d.	3.87
	904	83.3	13.3		3.3	2,500.0	400		100	1,000.0	def.	def.		n.d.	12.24
	1058	85.9	9.2		4.9	1,651.5	3,000		230.0	1,307.4	def.	def.		n.d.	3.58
	1211	98.0			2.0	2,332.6			201.4	1,918.2	def.			n.d.	3.75
		97.4	0.4		2.1	2,529.8	500		238.0	2,067.8	def.	def.		n.d.	3.97
Emanuel	49	92.6	0.6		6.8	482.5	225.0		230.5	446.2	def.	def.		n.d.	0.47
	50	94.1			5.9	686.5			232.5	615.9	def.			n.d.	0.48
	53	83.8			16.2	367.9			370.0	368.2	def.			n.d.	0.39
	57	91.5	1.7		6.8	430.9	183.3		223.5	397.0	def.	def.		n.d.	0.40
	58	90.6			9.4	682.7			349.0	626.2	def.			n.d.	0.75
	59	94.2			5.8	374.3			200.0	356.3	def.			n.d.	0.36
	395	92.8			7.2	394.7			188.5	366.1	def.			n.d.	0.42
	1208	97.1			2.9	392.9			121.4	368.9	def.			n.d.	0.42
		90.3	3.5		6.2	477.3	266.7		175.0	420.8	def.	def.		n.d.	0.66
Greene		95.3	1.3	3.4	n.d.	4,964.4	4,000.0	5,840.9	n.d.	4,595.8	def.	def.	def.	n.d.	6.78

County	ID	c1	c2	c3	c4	c5	c6	c7	c8	c9	c10	c11	c12	c13	c14
	316	95.0	0.2	n.d.	n.d.		343.6	1,308.7	n.d.	1,189.9	n.d.	def.		n.d.	4.93
	404	100.0		n.d.	n.d.			900.7	n.d.	722.0	def.			n.d.	4.25
	407	100.0		n.d.	n.d.			2,075.1	n.d.	1,632.9	5.16			n.d.	5.16
	408	100.0		n.d.	n.d.			2,014.0	n.d.	1,549.2	def.			n.d.	5.46
	444	99.2	0.8	n.d.	n.d.		306.7	1,390.6	n.d.	1,038.6	def.	def.		n.d.	5.09
	478	98.4	1.5	n.d.	n.d.		1,000	806.0	n.d.	673.7	def.	def.		n.d.	4.52
	544	100.0		n.d.	n.d.			1,841.3	n.d.	1,453.7	def.			n.d.	5.66
	550	100.0		n.d.	n.d.			1,451.3	n.d.	1,255.7	def.			n.d.	5.85
	562	100.0	0.0	n.d.	n.d.		35	2,063.4	n.d.	1,507.9	4.50	def.		n.d.	4.50
	571	100.0	0.2	n.d.	n.d.		350	1,117.8	n.d.	854.5	def.	def.		n.d.	3.93
		99.7		n.d.	n.d.			1,976.8	n.d.	1,554.0	5.25	5.83		n.d.	5.25
Habersham		98.0	0.2	1.8	n.d.	4,183.3	287.5	1,432.5	n.d.	901.7	3.27	2.42	2.64	n.d.	3.25
Harris		94.6	2.8	2.6	n.d.	3,179.7	4,558.5	2,846.0	n.d.	2,327.6	6.72	8.80	2.82	n.d.	6.53
	672	95.4	4.6		n.d.		10,000	3,514.2	n.d.	3,149.8	7.14	8.33	1.67	n.d.	7.19
	679	96.7	3.1		n.d.	200	2,437.5	2,790.5	n.d.	2,360.0	7.14	9.38		n.d.	7.16
	695	99.9		0.1	n.d.	175		2,715.9	n.d.	2,394.2	6.12		4.73	n.d.	6.12
	696	98.8		0.1	n.d.	800.0		3,217.7	n.d.	2,567.8	6.26		5.33	n.d.	6.25
	703	95.1		1.2	n.d.			2,254.8	n.d.	1,949.2	7.47		8.43	n.d.	7.51
	707	98.6		4.9	n.d.	7,000		2,301.7	n.d.	1,734.1	6.47	4.93		n.d.	6.44
	717	97.7	1.4	1.7	n.d.		850.0	3,002.7	n.d.	2,164.4	5.58	3.33	5.59	n.d.	5.56
	781	77.8	0.5	12.0	n.d.	1,650.0	1,000	2,474.1	n.d.	2,845.2	7.08	19.41	1.54	n.d.	5.17
	782	80.4	10.2		n.d.	9,750.0	16,500	2,938.9	n.d.	2,079.5	6.37	3.84	5.51	n.d.	6.14
	786	90.2	1.6	18.0	n.d.	2,728.6	863.5	3,032.6	n.d.	2,429.1	7.18	8.00		n.d.	7.26
	920	99.2	9.8		n.d.		17,200	2,657.1	n.d.	2,381.0	7.32	6.00		n.d.	7.30
	934	100.0	0.8		n.d.		1,200	6,586.2	n.d.	4,842.8	7.68			n.d.	7.68
	1186	99.2		0.8	n.d.	500	500	1,372.8	n.d.	864.9	5.74	2.50		n.d.	5.68
Houston		98.5		1.4	0.1	16,033.3		7,363.0	def.	6,586.8	9.58	9.58	10.42	n.d.	9.57
Thomas		88.4	3.8	7.8	n.d.	23,700.0	3,833.3	4,867.0	n.d.	4,659.8	6.72	4.65	9.38	n.d.	6.73
Towns		98.1	0.7	1.2	n.d.	1,610.0	340.0	1,134.5	n.d.	653.8	4.21	2.05	2.88	n.d.	4.16
Twiggs		96.1	0.0	3.8	0.1	19,416.7	n.a.	5,301.1	n.a.	3,668.8	6.56	5.00	10.00	n.d.	6.64
	323	87.5		12.5		15,750		2,694.9	n.a.	2,104.0	6.51		10.00	n.d.	6.81
	324	100.0			n.d.			11,539.5	n.a.	10,962.5	9.44			n.d.	9.44
	325	100.0			n.d.			11,566.3	n.a.	8,208.4	9.35			n.d.	9.35
	326	100.0			n.d.			4,371.3	n.a.	3,035.6	6.25			n.d.	6.25
	354	99.5	0.5		n.d.		500	4,144.3	n.d.	2,640.3	5.60	2.70		n.d.	5.58
	355	100.0			n.d.			2,846.2	n.d.	1,697.7	7.44			n.d.	7.44
	356	78.9		21.1	n.d.	35,000		5,450.7	n.a.	4,145.4	7.00		10.00	n.d.	7.47
	372	99.3		def.	n.d.	def.		5,453.0	n.d.	3,703.2	5.08	def.	def.	n.d.	5.08
	396	100.0		7.0	n.d.	7,500		11,373.8	n.a.	6,066.0	5.19		10.00	n.d.	5.19
	425	93.0			n.d.			6,187.0	n.d.	5,916.2	5.08			n.d.	5.26
Wilkes		94.4	3.6	1.9	n.d.	3,875.0	2,063.0	4,568.9	n.d.	4,213.3	def.	def.	def.	n.d.	5.49
Worth		72.6	1.6	12.0	13.8	5,661.1	772.7	1,781.2	1,384.6	1,819.5	3.39	2.74	5.78	5.06	3.73

(continued)

Appendix G (Continued)

County	Militia District	Personal Property (%)			Mean Personal Property ($)				% Producing Cotton					
		1a	3a	2a	1a	3a	2a	All Operators	1a	3a	6a	2a	All Operators
Baker	Milford	96.0	1.7	2.2	12,894.9	4,448.6	1,114.7	10,164.9	85.1	85.7	80.0	88.9	84.9
		95.4	0.1	4.5	12,823.0	133.3	1,416.7	8,959.8	81.0	66.7	78.9	88.9	81.6
	Newton	96.3	2.5	1.2	12,927.8	7,685.0	812.8	10,830.9	87.0	100.0	81.8	88.9	86.8
Baldwin	105	98.0	0.6	1.4	26,431.5	1,505.4	1,476.2	19,754.0	74.3	65.2	81.0	42.8	68.4
	115	99.0	0.1	0.9	11,324.3	250	625.0	9,530.7	90.0	100.0	100.0	40.0	83.8
	318	92.9	0.2	6.9	7,944.3	375.0	1,455.9	5,887.9	95.2	100.0	50.0	47.0	81.0
	319	97.6	1.0	1.5	79,359.8	6,891.7	6,205.0	62,207.5	80.8	100.0	100.0	33.3	79.5
	320	97.5	0.4	2.1	19,370.8	1,275.0	2,128.6	15,805.2	100.0	100.0	100.0	100.0	100.0
	321	100.0	0.0		29,230.9	0		26,795.0	21.7	0.0			19.2
	322	99.5	0.3	0.2	42,998.8	745.8	167.3	26,461.5	32.2	33.3	25.0	6.7	25.0
		95.7	3.8	0.5	6,521.3	846.4	157.0	4,477.1	78.3	71.4	100.0	60.0	75.0
Clay	431	97.8	0.3	1.8	7,290.8	1,356.2	584.5	5,960.8	91.8	100.0	100.0	88.2	91.4
		97.9		2.0	6,495.4		495.4	5,201.3	82.5		100.0	63.6	79.2
	749	98.3	0.5	1.2	9,735.2	1,766.7	555.6	7,981.2	96.3	100.0	100.0	91.3	95.6
	969	96.2	0.0	3.8	4,066.2	125	681.2	3,383.0	90.3	100.0	100.0	100.0	92.4
Coffee	437	95.0	0.3	4.7	3,015.6	576.8	388.2	2,269.4	34.4	50.0		10.9	28.2
	748	97.2	0.0	2.8	3,124.8	0	278.4	2,396.8	43.2	100.0		14.3	37.3
	1026	96.0		4.0	2,845.8		255.6	2,021.6	22.2			0.0	15.2
	1030	94.5	2.0	3.6	2,094.5	769.0	208.6	1,545.9	34.0	33.3		5.0	26.3
	1127	93.7		6.3	2,500.9		841.4	2,224.3	37.1			14.3	33.3
	1170	96.0		4.0	5,948.2		734.4	4,644.8	30.3			18.2	27.3
		88.9		11.1	2,141.5		437.2	1,493.8	41.9			21.0	34.0
Echols	719	93.0	0.0	7.0	3,730.9	0	1,013.8	3,060.4	67.8	66.7		36.0	61.0
	904	77.4	0.0	22.6	5,665	0	1,650	2,438.3	0.0	100.0		0.0	33.3
	1058	88.7	0.0	11.3	3,493.6	0	1,081.3	2,678.4	58.8	0.0		14.3	44.0
	1211	97.6		2.4	4,992.6		505.1	4,120.1	65.5			57.1	63.9
		90.9	0.0	9.1	2,928.9	0	1,258.9	2,565.4	74.4	100.0		40.0	68.5
Emanuel	49	96.1	0.3	3.6	2,939.8	750.0	702.0	2,618.8	64.9	50.0		50.7	62.9
	50	95.4		4.6	5,333.8		1,402.8	4,722.3	78.9			50.0	74.4
	53	94.7		5.3	2,408.0		698.0	2,132.2	53.8			60.0	54.8
	57	96.0		3.0	2,171.8	500.0	479.5	1,907.4	70.0	33.3		40.0	65.1
	58	94.3	0.9	5.7	5,099.8		1,510.1	4,491.4	69.4			50.0	66.1
	59	98.3		1.7	1,495.0		229.1	1,364.6	70.5			71.4	70.6
	395	98.4		1.5	2,188.2		213.5	1,914.0	69.4			50.0	66.7
	1208	99.0		1.0	2,144.8		215.7	1,973.9	29.2			14.3	27.8
		95.6	2.9	1.5	2,283.3	1,000.0	196.2	1,902.8	79.1	66.7		75.0	77.8
Greene		97.8	0.8	1.3	14,156.8	7,100.0	2,253.7	13,125.9	92.0	100.0	90.9	93.5	92.2

230

(continued)											
	316	94.9	0.1	2.0		933.5	163.8	201.6	2,231.1 788.3	37.3	69.4 88.0	76.3 90.5
	404	95.3	0.0	5.1		3,737.8		724.3	3,125.7	91.1	61.5	71.9
	407	98.8		4.7		4,836.6		206.5	3,832.5	74.5	61.1	71.1
	408	95.5	0.3	1.2		2,931.9	223.3	427.3	2,288.9	73.8	88.0	90.0
	444	94.5	.2	4.2		635.5	100	177.7	554.5	90.2	37.5	54.6
	478	97.2		5.3		4,215.5		456.6	3,424.1	58.8	95.0	90.5
	544	98.8		2.8		2,471.9		187.8	2,164.4	89.3	57.1	80.8
	550	96.7		1.2		3,428.6	225.0	324.5	2,613.8	84.4	42.8	37.5
	562	97.8	.2	3.2		1,982.8	250	143.7	1,567.4	35.6 0.0	55.6	72.6
	571	98.4	.1	1.9		3,551.8		208.9	2,830.5	78.8 0.0	88.9	92.0
Habersham		93.8	0.0	6.1		1,976.2	50.0	216.1	1,310.7	94.2 0.0	4.8	4.6
Harris	672	95.1	0.1	4.8		10,228.0	462.1	2,091.5	8,498.1	4.6 94.4	79.5	89.6
	679	97.4	0.0	2.6		12,414.7	n.a.	2,200.9	10,924.1	91.9	55.6	75.7
	695	97.1	0.0	2.9		9,943.9	n.a.	1,783.3	8,508.0	78.3 100.0	100.0	98.5
	696	93.0		7.0		12,513.0		7,903.8	12,024.2	98.1	85.7	92.5
	703	88.1		11.8		10,250.4		6,436.3	9,215.9	94.9 100.0	88.9	92.4
	707	90.0		10.0		6,704.8		3,732.7	6,209.4	92.8 100.0	91.7	95.9
	717	97.3	0.0	2.7		7,131.4	n.a.	659.4	5,495.0	96.7	68.8	83.3
	781	94.5	0.6	4.9		11,194.9	4,465	1,556.2	8,538.8	88.9 50.0	69.6	86.5
	782	99.6	0.0	0.3		6,567.6		383.3	6,110.9	91.9 100.0	100.0	94.7
	786	92.0	0.0	8.0		11,825.4	n.a.	2,280.4	8,467.8	94.1 100.0	76.9	92.2
	920	95.8	0.0	4.2		10,731.5	n.a.	1,335.8	8,174.1	96.6 100.0	94.7	91.9
	934	99.2	0.0	0.8		10,256.5	n.a.	758.3	9,205.8	90.7 100.0	100.0	95.3
	1186	97.4 89.4	0.6	2.6 9.9		29,926.6 3,253.7	1,080	2,176.1 622.3	22,580.8 2,273.3	94.7 92.0 100.0 86.7	44.4 80.8	79.4 84.7
Houston		96.1		3.9		16,163.8		5,497.1	15,027.0	87.5 66.7	60.0	84.4
Thomas		97.6	1.0	1.3		12,329.8	2,528.0	1,494.8	10,870.8	86.3 100.0	93.5	87.2
Towns		87.9	0.1	12.0		845.8	60.0	163.6	557.3	0.0 0.0	0.0	0.0
Twiggs	323	95.8	0.0	4.2		14,923.8	n.a.	1,405.9	10,624.6	90.1 100.0	70.6	84.1
	324	94.6		5.3		6,335.2		814.2	4,650.9	87.8 100.0	55.6	78.3
	325	99.4		0.6		30,086.6		3,295	28,747.0	78.9	0.0	75.0
	326	93.5		6.5		32,072.4		5,935.9	24,944.2	87.5	55.6	78.8
	354	98.2		1.8		13,707.5		598.8	9,810.4	88.5	72.7	83.8
	355	98.6	0.1	1.3		9,959.8	427	274.8	6,800.4	93.3 100.0	78.6	88.9
	356	92.1		7.9		5,216.2		669.9	3,397.7	95.6	76.1	87.8
	372	94.1		5.8		15,593.9		1,615.1	10,351.8	100.0 100.0	86.7	95.1
	396	95.3		4.6		14,363.3		1,501.1	10,270.8	96.7 100.0	71.4	88.9
	425	95.5 98.5		4.5 1.5		53,701.2 18,941.4		3,977.0 5,500	34,364.0 18,269.4	72.7 73.7 100.0	57.1 0.0	66.7 71.4
Wilkes		98.6	1.2	0.2		20,785.4	3,004.5	1,103.8	18,770.1	95.2 87.5	92.3	95.0
Worth		89.9	1.2	8.9		2,582.1	637.3	1,045.4	2,213.7	70.3 90.9	78.8	72.7

(continued)

Appendix G (Continued)

County	Militia District		% Bales/Category			Mean Bales					Improved Acres/Bale (All Operators)				
		1a	3a	6a	2a	1a	3a	6a	2a	All Operators	1a	3a	6a	2a	All Operators
Baker	Milford	55.7	0.4	41.9	2.0	42.9	6.7	131.3	5.4	48.8	6.7	8.5	6.3	8.8	6.6
		53.8	0.4	42.8	3.1	43.2	5.0	116.8	5.3	44.1	6.8	12.5	7.7	11.2	7.3
	Newton	56.6	0.5	41.5	1.4	42.8	7.5	139.4	5.5	51.2	6.6	7.2	5.7	6.9	6.2
Baldwin	105	70.1	1.5	24.2	4.2	26.1	5.9	82.8	10.1	27.4	7.2	7.5	5.0	7.3	6.6
	115	95.5	0.6	1.5	2.4	19.0	3	8	6.5	17.3	8.6	6.7	10.0	13.3	8.7
	318	80.7	0.4	14.6	4.4	15.2	1.5	110	4.1	14.8	8.0	8.3	7.1	14.4	8.1
	319	45.8	1.9	50.0	2.2	48.0	14.0	122.2	24.0	62.8	6.7	6.1	4.5	5.3	5.5
	320	80.5	0.5	10.1	8.8	33.3	4.0	37.0	16.0	29.8	5.4	5.8	4.9	3.8	5.2
		100.0	0.0			40.2	0			40.2	10.6	n.d.			10.6
	321	93.6	1.9	0.6	3.8	29.2	3.0	2	12	22.3	9.0	17.8	152.5	22.2	10.6
	322	78.8	7.4	11.5	2.3	15.3	5.2	40	2.7	12.9	7.3	7.6	6.5	13.8	7.4
Clay	431	84.7	0.3	9.8	5.2	22.2	4.0	86.3	6.1	20.6	def.	def.	def.	n.d.	7.6
	749	81.9		8.8	9.3	23.3		41.5	12.4	22.4	def.	def.	def.	n.d.	8.3
		84.7	0.4	11.2	3.7	27.3	4.7	187.5	5.9	25.8	def.	def.	def.	n.d.	6.9
	969	87.4	0.2	6.0	6.4	13.4	2	30.0	3.8	11.7	def.	def.	def.	n.d.	9.1
Coffee	437	94.0	1.7		4.3	5.3	4.0		2.0	4.9	25.0	26.0		58.1	26.4
	748	90.3	6.8		2.9	4.9	7		1.5	4.7	18.6	18.6		69.7	20.1
		100.0			0.0	1.3			n.a.	1.3	130.9			n.a.	149.0
	1026	96.4	1.8		1.8	3.0	1		1	2.8	44.3	78.0		234.0	48.3
	1030	97.0			3.0	2.5			1	2.4	40.5			82.0	4.7
	1127	95.3			4.7	22.2			5.5	19.4	12.2			14.3	12.3
	1170	87.1			12.9	2.1			1.0	1.8	44.7			61.2	46.8
Echols	719	94.6	1.3		4.0	4.6	2.0		1.3	4.2	21.7	42.5		66.1	23.8
		0.0	100.0		0.0	0	2		0	2	n.a.	30.0		n.a.	67.5
	904	97.5	0.0		2.5	7.7	0		7.2	7.2	15.7	n.a.		80.5	18.2
	1058	94.4			5.6	5.4			1.5	4.7	21.2			61.0	23.4
	1211	94.5	1.8		3.6	3.2	2		1.0	3.0	26.1	20.0		64.0	27.4
Emanuel	49	91.4	0.5		8.0	3.5	2.0		2.5	3.4	def.	def.		n.d.	34.4
	50	94.8			5.2	4.2			2.0	3.0	def.			n.d.	33.7
	53	72.3			27.6	2.4			4.3	2.8	def.	def.		n.d.	27.9
	57	93.1	0.7		6.2	2.8	1		2.2	2.7	def.			n.d.	41.8
	58	89.9			10.1	5.5			4.2	5.3	def.			n.d.	28.6
	59	90.5			9.5	2.4			2.2	2.4	def.			n.d.	35.5
	395	92.3			7.7	2.8			2.0	2.7	def.			n.d.	31.4
	1208	98.3	3.1		1.7	2.7	2.5		1	2.6	def.	def.		n.d.	80.2
		89.5			7.4	4.3			2.0	3.8	def.			n.d.	23.2
Greene		91.5	1.9	4.3	2.3	23.5	27.2	36.6	6.8	22.6	def.	def.	def.	def.	14.0

232

County	#													
Gwinnett	316	88.5	0.4		11.1	3.6			2.0	3.3	def.	def.	def.	26.1
	404	82.5			17.5	2.6			2.3		def.	def.		18.4
	407	90.5			9.5	3.5			1.8		def.	def.		30.1
	408	88.5			11.5	3.7			2.1		def.	39.6		37.7
	444	85.3	2.8		11.9	4.1		3.3	2.0		23.4	def.	def.	22.3
	478	94.3	0.0		5.7	2.1		n.a.	1.0		def.	37.8	n.a.	39.7
	544	85.8			14.2	4.6			2.7		71.2	def.		19.2
	550	95.2			4.8	2.6			1.2		n.d.	32.9		32.4
	562	79.0	0.0		21.0	2.3			1.4		22.4	103.3		81.6
	571	89.2	0.0		10.8	2.8		n.a.	1.8		n.d.	29.6	n.a.	29.0
		91.9	0.0		8.1	5.9		n.a.	2.1		23.8	23.2	n.a.	21.4
Habersham		76.2	0.0	0.0	23.8	2.2		n.a.	1.1		n.d.	def.	n.a.	501.8
Harris	672	88.1	1.9	3.7	6.3	21.9	25.7	32.2	7.3	19.7	n.d.	def.	def.	10.5
	679	94.7	3.4		1.8	26.4	45		4.8	24.7	n.d.	def.	def.	11.4
	695	89.2	4.0	0.4	6.5	23.0	27.0	5	9.9	21.0	n.d.	def.	def.	9.8
	696	93.6		0.0	6.4	25.7		n.a.	16.5	24.8	n.d.	def.	n.a.	11.4
	703	92.8		0.6	6.6	20.6		2.5	7.1	17.7	n.d.	def.	def.	14.3
	707	81.6		9.5	8.9	13.3		45.0	7.6	13.3	n.d.	def.	def.	11.2
	717	90.6	0.2		9.1	14.9	2	16.7	6.5	13.2	n.d.	def.	def.	10.5
	781	86.0	0.4	3.6	10.0	21.0	5	90.5	8.8	18.1	n.d.	def.	def.	10.1
	782	78.2	2.7	16.2	3.0	18.2	30	31.0	11.0	20.7	n.d.	def.	def.	10.0
	786	70.5	3.6	21.2	4.6	25.8	18.5		4.7	21.7	n.d.	def.	def.	9.1
	920	86.2	6.4		7.4	27.3	100		6.4	22.9	n.d.	def.		8.8
	934	97.2	0.4		2.4	21.4	5		4.7	19.5	n.d.	def.		10.2
	1186	96.4			3.6	53.4			11.5	47.2	n.d.	def.		11.3
		79.9	0.9		19.1	11.0	5		4.9	8.8	n.d.	def.	def.	9.2
Houston		96.9		0.8	2.2	63.6		124.0	17.6	60.4	n.d.	def.	def.	6.3
Thomas		84.7	4.9	5.6	4.8	22.7	24.8	73.4	11.0	22.5	def.	def.	def.	11.6
Towns		n.a.	n.a.	n.a.	n.a.	n.a.	n.a.	n.a.	n.a.	n.a.	n.a.	n.a.	n.a.	n.a.
Twiggs	323	90.3	0.0	3.8	5.8	46.0		129.8	8.2	36.9	n.d.	def.	def.	7.6
	324	69.3		22.9	7.8	19.8	n.a.	236	8.1	21.9	def.	8.4	def.	8.2
	325	100.0			0.0	100.7			n.a.	100.7	n.d.	def.		8.4
	326	95.9			4.1	98.2			17.6	82.7	n.d.	def.		7.3
	354	94.0			5.9	31.6			5.8	25.0	n.d.	def.		9.6
	355	95.3	0.3		4.4	36.5	3		4.3	26.8	n.d.	def.		8.4
	356	87.0		13.6	13.0	21.6		230	6.0	16.2	n.d.	def.	def.	7.2
	372	79.5		0.6	6.9	53.6		8	8.9	43.2	n.d.	def.	def.	6.3
	396	91.2			8.2	39.8			10.4	31.6	n.d.	def.		7.4
	425	93.8		4.9	6.2	171.5		45	22.8	121.9	n.d.	8.7		6.8
		95.1			0.0	61.8			n.a.	60.7	n.a.		11.1	8.8
Wilkes		93.2	3.2	3.0	0.7	24.9	10.0	36.4	4.7	23.4	17.5	15.4	16.7	11.2	15.4
Worth		65.9	1.4	21.8	10.9	7.4	3.0	36.3	4.4	8.0	14.9	14.9	18.0	8.3	13.5

(*continued*)

Appendix G (Continued)

County	Militia District	Improved Acres/Bale (Producers Only)				All Operators
		1a	3a	6a	2a	
Baker	Milford	6.4	7.6	5.5	8.2	6.1
	Newton	6.7	9.0	6.3	10.6	6.6
		6.3	7.2	5.2	6.3	5.8
Baldwin	105	6.5	6.4	4.6	4.9	6.0
	115	8.0	6.7	10.0	8.5	8.1
	318	7.8	8.3	4.4	9.6	7.4
	319	6.2	6.1	4.5	4.0	5.3
	320	5.4	5.8	4.9	3.8	5.2
	321	7.7	n.d.			7.7
	322	5.9	6.7	10.0	2.9	5.8
		6.4	7.0	6.5	6.9	6.5
Clay	431	def.	def.	def.	n.d.	7.4
	749	def.	def.	def.	n.d.	8.0
	969	def.	def.	def.	n.d.	6.8
		def.	def.	def.	n.d.	8.6
Coffee	437	13.9	20.0		13.8	14.0
	748	13.0	18.6		12.7	13.3
	1026	32.2			n.a.	32.2
	1030	24.4	30.0		38.0	24.7
	1127	25.0			25.0	25.0
	1170	7.7			8.3	7.7
		25.2			21.2	24.7
Echols	719	16.5	25.0		16.2	16.6
	904	n.a.	30.0		n.a.	30.0
	1058	13.2	n.a.		15.0	13.2
	1211	15.8			14.2	15.7
		19.8	20.0		20.0	19.8
Emanuel	49	def.	def.		n.d.	26.6
	50	def.			n.d.	28.6
	53	def.			n.d.	20.2
	57	def.	def.		n.d.	33.2
	58	def.			n.d.	24.7
	59	def.			n.d.	27.8
	395	def.			n.d.	26.1
	1208	def.	def.		n.d.	33.4
					n.d.	19.0
Greene		def.	def.	def.	def.	13.4

Gwinnett	316	def.	det.	def.	22.1
	404	def.		n.d.	17.2
	407	34.1		n.d.	25.8
	408	def.	def.	15.3	32.0
	444	24.6	n.a.	def.	21.3
	478	def.		26.8	24.7
	544	29.5		n.d.	17.3
	550	42.6		14.4	28.8
	562	25.7	n.a.	n.d.	33.7
	571	22.5	n.a.	13.6	24.4
				n.d.	20.7
Habersham		def.	n.a.	n.a.	48.6
Harris	672	def.	def.	n.d.	10.3
	679	def.	def.	n.d.	11.2
	695	def.	def.	n.d.	9.6
	696	def.	n.a.	n.d.	11.2
	703	def.	def.	n.d.	13.6
	707	def.	def.	n.d.	10.9
	717	def.	def.	n.d.	9.9
	781	def.	def.	n.d.	10.0
	782	def.	def.	n.d.	9.4
	786	def.	def.	n.d.	9.1
	920	def.	def.	n.d.	8.6
	934	def.		n.d.	10.1
	1186	def.	def.	n.d.	11.3
				n.d.	8.5
Houston		def.	def.	n.d.	6.1
Thomas		def.	def.	def.	10.9
Towns		n.a.	n.a.	n.a.	n.a.
Twiggs	323	def.	def.	n.d.	7.5
	324	8.0	def.	def.	7.8
	325	def.		n.a.	8.0
	326	def.		n.d.	7.2
	354	def.	def.	n.d.	9.4
	355	def.		n.d.	8.4
	356	def.	def.	n.d.	7.2
	372	def.	def.	n.d.	6.3
	396	def.		n.d.	7.4
	425	8.5	11.1	n.a.	6.6
					8.6
Wilkes		15.1	16.0	11.0	17.3	15.0
Worth		12.5	16.4	7.8	13.7	11.7

(*continued*)

235

Appendix G (Continued)

Notes: We assume acres that are not assigned to tenants are assigned to proprietors. Thus, percentage improved of total acres is recorded as defective if only improved or unimproved acres are not defined; if both are defined, or if both are not defined, values are calculated. We assume farm value that is not assigned to tenants is assigned to proprietors. Thus, farm value/total acre is recorded as defective if only improved or unimproved acres are not defined; if both are defined, or if both are not defined, values are calculated.

BAKER COUNTY

Milford District:

Category 1a contains two records defective for improved and unimproved acres, producing no cotton, and four records defective for farm value, two of which report 152 improved acres and no unimproved acres.

Category 2a contains three records defective for improved acres, producing 15 bales of cotton.

Category 6a contains one record defective for improved and unimproved acres and farm value, producing 100 bales of cotton.

Newton District:

Category 1a contains seven records defective for farm value, reporting 837 improved acres and no unimproved acres.

Category 2a contains one record defective for improved acres, producing 1 bale of cotton.

BALDWIN COUNTY

District 105:

Category 3a contains one record designated "renter" on schedule IV and assigned no farm value but assigned real property on schedule I.

District 115:

Category 3a contains two records designated "renter" on schedule IV and assigned no farm value but assigned real property on schedule I. Page 19 of schedule IV for Baldwin County (which is the first page for this district) is missing in the manuscript returns, resulting in a loss of approximately forty farms. The corresponding pages on schedule I (pages 91–93, line 33) were therefore not coded.

District 318:

Category 2a contains one record designated "renter" on schedule IV that cannot be located on schedule I.

Category 3a contains one record designated "renter" on schedule IV and assigned no farm value but assigned real property on schedule I. This record reports 115 improved acres, no unimproved acres, and produces 27 bales of cotton.

District 319:

Category 2a contains one record designated "renter" on schedule IV that cannot be located on schedule I.

Category 3a contains one record designated "renter" on schedule IV and assigned no farm value but assigned real property on schedule I. This record reports 16 improved acres, no unimproved acres, and produces 2 bales of cotton.

District 320:

Category 1a contains one public institution not included in the calculation of personal property.

District 321:

Category 1a contains one public institution not included in the calculation of personal property.

Category 2a contains two records designated "renter" on schedule IV that cannot be located on schedule I.

Category 3a contains three records designated "renter" on schedule IV and assigned no farm value but assigned real property on schedule I. These records report 65 improved acres, no unimproved acres, and produce 2 bales of cotton.

CLAY COUNTY

District 749:

Category 3a contains two records designated "renter" on schedule I and defective for improved and unimproved acres and farm value but assigned real property on schedule I. These two records produce 11 bales of cotton.

District 969:

Category 1a contains three records defective for improved acres, producing 53 bales of cotton, and reporting 875 unimproved acres and $2,450 farm value.

Category 3a contains one record designated "renter" on schedule IV and defective for improved and unimproved acres and farm value but assigned real property on schedule I.

COFFEE COUNTY

District 437:

Category 1a contains one record defective for improved acres and unimproved acres, producing no cotton, and one record defective for farm value, reporting 20 improved acres and no unimproved acres.

Category 2a contains one record defective for improved acres, producing no bales of cotton.

District 748:

Category 1a contains one record defective for improved acres and unimproved acres, producing no bales of cotton, and two records defective for farm value, one of which reports 60 improved acres and no unimproved acres.

Category 2a contains one record defective for improved acres, producing no bales of cotton.

District 1026:

Category 1a contains one record defective for improved acres and unimproved acres, producing no bales of cotton, and two records defective for farm value, one of which reports 8 improved acres and no unimproved acres.

Category 2a contains four records defective for improved acres, producing no bales of cotton.

District 1030:

Category 1a contains one record defective for improved acres, unimproved acres and farm value, producing no bales of cotton.

Category 2a contains two records defective for improved acres, producing no bales of cotton.

District 1127:

Category 2a contains two records defective for improved acres, producing no bales of cotton.

District 1170:

Category 2a contains two records defective for improved acres, producing no bales of cotton. Only one record in this category reports unimproved acres in this district. The remaining records in this category are not subtracted from county total producers for computing mean unimproved acres because we assume their unimproved acres are included in category 1. Unlike the Gwinnett County return, the return for Coffee seems highly consistent with this single exception.

EMANUEL COUNTY

District 53:

Category 1a contains two records defective for improved acres, one of which produces 2 bales of cotton, and one record defective for unimproved acres. These records report a combined total of 400 unimproved acres and $300 farm value.

(*continued*)

237

Appendix G (Continued)

District 57:

Category 1a contains two records defective for improved acres and unimproved acres, one of which produces 2 bales of cotton, and reporting a combined total of $300 farm value.

District 58:

Category 1a contains one record defective for improved acres and unimproved acres, producing 2 bales of cotton, and reporting $200 farm value.

District 395:

Category 1a contains one record defective for improved acres and unimproved acres, producing no bales of cotton, and reporting $150 farm value.

GREENE COUNTY

Category 1a has three records defective for improved acres, unimproved acres, and farm value, producing 11 bales of cotton.

Category 2a contains only six records reporting improved acres, four of which produce 23 bales of cotton. Improved acres/bale is therefore not computed for this category. The remaining records in this category are not subtracted from county total producers in computing mean improved acres because we assume their improved acres are included in categories 1, 3, or 6.

GWINNETT COUNTY

County Level:

In category 2a improved acres is consistently defined in only four districts. We are therefore reporting the acreage but are not computing the mean values, cotton/improved acre, and farm value/total acre for this category. Similarly, unimproved acres in this category are consistently reported in no districts, and we are therefore following comparable procedures.

District 404:

Category 1a contains three records defective for improved acres, unimproved acres, and farm value, one of which produces 1 bale of cotton.

In collating schedules I and IV we encountered six contiguous names on schedule I who were "farmers" and heads of household with or without real property, or who were "farmers" not heading a household who held real property, all of whom were missing from schedule IV. These names were found on page 184 of schedule I. We consider schedule IV defective for these names, and schedule I data for these names have been excluded from the statistics.

District 405:

In collating schedules I and IV we encountered eighty-nine contiguous names on schedule I who were "farmers" and heads of household with or without real property, or who were "farmers" not heading a household who held real property, all of whom were missing from schedule IV. These names were found on pages 239–255 of schedule I. We consider schedule IV defective for these names, and schedule I data for these names have been excluded from the statistics.

District 406:

In collating schedules I and IV we encountered sixty-nine contiguous names on schedule I who were "farmers" and heads of household with or without real property, or who were "farmers" not heading a household who held real property, all of whom were missing from schedule IV. These names were found on pages 176–183 and pages 185–191 of schedule I. We consider schedule IV defective for these names, and schedule I data for these names have been excluded from the statistics.

District 407:

Category 1a contains five records defective for farm value, reporting 200 improved acres and no unimproved acres.

District 408:

Category 1a contains two records defective for improved acres, unimproved acres, and farm value, one of which produces 1 bale of cotton, and reporting 50 unimproved acres and $250 farm value.

Category 1a contains one record defective for improved acres, producing 1 bale of cotton, and one record defective for improved acres.

Category 2a has only two records reporting improved acreage, one of which produces 2 bales of cotton. Improved acres/bale is therefore not computed for this category. The remaining records in this category are not subtracted from county total producers for computing mean improved acres because we assume their improved acres are included in categories 1 or 3.

In collating schedules I and IV we encountered five contiguous names on schedule I who were "farmers" and heads of household with or without real property, or who were "farmers" not heading a household who held real property, all of whom were missing from schedule IV. These names were found on page 192 of schedule I. We consider schedule IV defective for these names, and schedule I data for these names have been excluded from the statistics.

District 444:

Category 1a contains one record defective for improved acres, unimproved acres, and farm value, producing no bales of cotton.

Category 2a contains two records defective for improved acres, one of which produces 1 bale of cotton.

District 478:

Category 2a contains only one record reporting improved acres, producing no cotton. The remaining records in this category are not subtracted from county total producers for computing mean improved acres, since we assume their improved acres are included in category 1.

District 544:

Category 2a contains one record defective for improved acres, producing no cotton. This category also contains two farms reporting unimproved acreage. The remaining records in this category are not subtracted from county total producers for computing mean unimproved acres because we assume their unimproved acres are included in category 1.

In collating schedules I and IV we encountered seventeen contiguous names on schedule I who were "farmers" and heads of household with or without real property, or who were "farmers" not heading a household who held real property, all of whom were missing from schedule IV. These names were found on pages 172–175 of schedule I. We consider schedule IV defective for these names, and schedule I data for these names have been excluded from the statistics.

District 550:

Category 1a contains two records defective for improved acres, unimproved acres, and farm value, and which produce no bales of cotton.

District 562:

Category 1a contains five records defective for farm value, reporting 143 improved acres and 88 unimproved acres.

In category 2a only three farms report unimproved acreage. The remaining records in this category are not subtracted from county total producers for computing mean unimproved acres, since we assume their unimproved acres are included in category 1.

Category 3a contains one record designated "tenant" on schedule IV but with farm value absent and with real property recorded on schedule I. This record reports 15 improved acres and no unimproved acres.

District 571:

Category 1a has one record defective for improved acres, producing 1 bale of cotton and reporting 100 unimproved acres and $100 farm value.

In collating schedules I and IV we encountered thirty-one contiguous names on schedule I who were "farmers" and heads of household with or without real property, or who were "farmers" not heading a household who held real property, all of whom were missing from schedule IV. These names were found on pages 233–238 of schedule I. We consider schedule IV defective for these names, and schedule I data for these names have been excluded from the statistics.

HABERSHAM COUNTY

Category 1a contains six records defective for improved acres, producing no bales of cotton; seven records defective for unimproved acres; and six records defective for farm value. One of the latter reports 100 improved acres and $2,000 farm value, the unimproved acres being illegible.

Category 3a contains one record with "tenant" indicated in the occupation column of schedule I that reports real property and is also assigned farm value on schedule IV.

(*continued*)

Appendix G (Continued)

Bales of cotton in this county were commonly reported in fractional amounts. In our investigation we have encountered only one other reporting of a fractional bale, i.e., one report of one-half bale in Worth County. To retain strict apparent comparability, we have treated all bales with a reported value of less than one as 1 bale and all remaining fractional values to the nearest whole number. This procedure results in a total revised cotton production for Habersham County of 63 bales, which is 7.5 percent greater than the actual reported production of 58.58 bales.

HARRIS COUNTY

District 672:

Category 1a has one record defective for improved acres, unimproved acres, and farm value, producing 2 bales of cotton.

District 696:

Category 1a has one record defective for improved acres, unimproved acres, and farm value, producing 2 bales of cotton.

District 703:

Category 1a has one record defective for improved acres, unimproved acres, and farm value, producing 2 bales of cotton.

District 707:

Category 6a has one record defective for improved acres, unimproved acres, and farm value, producing 22 bales of cotton.

District 717:

Category 1a has two records defective for improved acres, unimproved acres, and farm value, one of which produces 5 bales of cotton.

District 786:

Category 6a has one record defective for improved acres, unimproved acres, and farm value, producing 11 bales of cotton.

District 920:

Category 1a contains two records defective for improved acres, unimproved acres, and farm value, one of which produces 7 bales of cotton.

Category 1a contains one record defective for improved acres, unimproved acres, and farm value, producing 12 bales of cotton.

HOUSTON COUNTY

Category 1a contains thirty-three records defective for improved acres, five of which produce 81 bales of cotton. This category also contains thirty records defective for unimproved acres and thirty-five records defective for farm value. Records which have defective values in less than three of the above variables report 555 improved acres, 2,002 unimproved acres, and $9,275 farm value. Three records defective for improved acres report 1,085 unimproved acres.

Category 2a contains four records that report farm value only, three of which produce 27 bales of cotton.

Category 3a contains four records defective for improved acres and unimproved acres, three of which produce 27 bales of cotton. The four records report $12,935 farm value.

THOMAS COUNTY

Category 1a contains nine records defective for improved acres, seven of which produce 299 bales of cotton. In addition, category 1a contains eight records defective for unimproved acres and ten records defective for farm value; these defective records report 1,190 improved acres, 1,625 unimproved acres, and $25,200 farm value.

Category 2a contains only one record that reports acreage; it reports 500 improved acres, 500 unimproved acres, and produces 9 bales of cotton.

TOWNS COUNTY

Category 1a contains ten records defective for improved acres, unimproved acres, and farm value, none of which produce cotton.

TWIGGS COUNTY

County Level:

In category 2a only one record reports improved acres. We are therefore reporting the acres but are not computing the mean values or improved acres/bale.

in category 6a there are three records in districts 323, 356, and 425, respectively, that report the name of the farm owner and his real and personal property but do not report sex, age, occupation, or other variables on schedule I. This pattern was also observed in Liberty County.

District 323:

Category 2a contains only one record reporting improved acres, producing 8 bales of cotton. Improved acres/bale is therefore not computed for this category, and the remaining records are not subtracted from county total producers for computing mean improved acres because we assume their improved acres are included in category 1.

District 325:

Category 1a contains three records defective for improved acres, one of which produces 48 bales of cotton; two records defective for unimproved acres; and two records defective for farm value, one of which reports 1,500 unimproved acres and is also defective for improved acres. One record defective for improved and unimproved acres reports $800 farm value.

District 326:

Category 1a contains one record defective for improved acres, unimproved acres, and farm value, producing 3 bales of cotton.

District 354:

Category 1a contains four records defective for improved acres, unimproved acres, and farm value, three of which produce 7 bales of cotton.

District 355:

Category 1a contains two records defective for improved acres, producing 9 bales of cotton, and one record defective for unimproved acres and farm value. One record defective only for improved acres reports 68 unimproved acres and $476 farm value.

District 356:

Category 1a contains one record defective for improved acres, unimproved acres, and farm value, producing 8 bales of cotton and reporting 200 unimproved acres.

District 372:

Category 1a contains one record defective for improved acres, unimproved acres, and farm value, producing no bales of cotton.

Category 2a contains one record reporting farm value only, producing 4 bales of cotton.

Category 6a contains one record defective for improved acres, unimproved acres, and farm value, producing 8 bales of cotton.

District 396:

Category 1a contains three records defective for improved acres, unimproved acres, and farm value, one of which produces 28 bales of cotton.

District 425:

Category 1a contains two records defective for improved acres and unimproved acres, producing no bales of cotton, and three records defective for farm value, one of which reports 4 improved acres.

WILKES COUNTY

Category 1a contains two records defective for farm value, reporting 85 improved acres and 110 unimproved acres.

Category 2a contains two records defective for improved acres and unimproved acres, producing 2 bales of cotton.

Category 3a contains one record defective for improved acres and unimproved acres, producing 10 bales of cotton and reporting $2,000 farm value.

WORTH COUNTY

Category 1a contains one record reporting one-half bale of cotton. This is the only fractional bale reported in the county and has been rounded to 1 bale.

Category 3a contains eight records designated "tenant" on schedule IV (as are all 2a records) that report real property on schedule I. In all cases the amount of real property reported is significantly less than the farm value reported on schedule IV.

Appendix H

STATISTICS BY MILITIA DISTRICT, TYPES 4 AND 5, SIXTEEN SAMPLE COUNTIES, 1860

County	Militia District	Type	No.	Total Personal Property ($)	Mean Personal Property ($)	Stand. Dev. Personal Property ($)	Total Real Property ($)	Mean Real Property ($)	Stand. Dev. Real Property ($)
Baker		4	13	44,140	3,395.4		28,690	2,206.9	
		5	32	67,350	2,104.7				
	Milford	4	1	1,300	n.a.	n.a.	300	n.a.	n.a.
		5	8	24,050	3,006.2	3,986.6			
	Newton	4	12	42,840	3,570.0	5,553.4	28,390	2,365.8	2,126.6
		5	24	43,300	1,804.2	2,846.7			
Baldwin		4	25	993,190	39,727.6		336,150	13,446.0	
		5	7	2,075	296.4				
	105	4	0						
		5	2	300	150.0	70.7			
	115	4	3	2,550	850.0	1,169.4	3,350	1,116.7	875.1
		5	2	275	137.5	17.7			
	318	4	2	10,150	5,075.0	6,965.0	3,300	1,650.0	1,202.1
		5	0						
	319	4	2	14,135	7,067.5	1,085.4	9,900	4,950.0	5,586.1
		5	1	1,000	n.a.	n.a.			
	320	4	6	251,915	41,985.8	30,642.1	86,000	14,333.3	7,257.2
		5	0						
	321	4	10	710,500	71,050.0	71,310.3	232,100	23,210.0	28,513.1
		5	2	500	250.0	70.7			
	322	4	2	3,940	1,970.0	2,644.6	1,500	750.0	353.6
		5	0						
Clay		4	26	309,576	11,906.8		85,100	3,273.1	
		5	24	25,746	1,072.8				
	431	4	9	289,650	32,183.3	42,146.8	73,400	8,155.6	11,046.2
		5	3	825	275.0	204.6			
	749	4	10	15,675	1,567.5	4,199.2	8,000	800.0	852.8
		5	10	3,600	360.0	728.3			

969	4	7	4,251	607.3	632.2	3,700	528.6	298.4
	5	11	21,321	1,938.3	4,370.8		
Coffee							
437	4	4	14,655	3,663.8		3,300	825.0
	5	7	1,300	185.7		
748	4	3	14,340	4,780.0	5,780.1	2,900	966.7	472.6
	5	1	280	n.a.	n.a.
1026	4	0			
	5	0	
1030	4	0	
	5	0
1127	4	2	275	137.5	88.4	400	n.a.	n.a.
	5	1	315	n.a.	n.a.
	4	3	80	26.7	2.9
1170	4	0			
	5	1	665	n.a.	n.a.
Echols					
	4	0
	5	0
Emanuel	
49	4	14	6,330	452.1	2,015.2	6,002	428.7	353.6
	5	22	3,178	144.4	70.7
	4	2	3,150	1,575.0
50	5	2	100	50.0
	4	0
53	5	1	100	n.a.	n.a.	1,488	248.0	220.5
	4	6	1,480	246.7	283.8
	5	3	530	176.7	136.5
57	4	0
	5	5	1,300	260.0	421.9
58	4	4	1,275	318.8	322.3	2,360	590.0	376.3
	5	5	589	117.8	119.7
59	4	1	290	n.a.	n.a.	304	n.a.	n.a.
	5	0
395	4	1	135	n.a.	n.a.	350	n.a.	n.a.
	5	3	279	93.0	95.8
1208	4	0
	5	3	280	93.3	121.0

(continued)

Appendix H (Continued)

County	Militia District	Type	No.	Total Personal Property ($)	Mean Personal Property ($)	Stand. Dev. Personal Property ($)	Total Real Property ($)	Mean Real Property ($)	Stand. Dev. Real Property ($)
Greene		4	25	136,270	5,450.8	5,776.0	100,355	4,014.2	4,995.0
		5	43	85,025	1,977.3	4,359.1			
Gwinnett									
	316	4	86	36,743	427.2		49,923	580.5	
		5	123	15,703	127.7				
	404	4	13	1,640	126.2	103.5	4,640	356.9	151.6
		5	16	1,337	83.6	100.6			
	405	4	6	3,790	631.7	765.0	4,400	733.3	294.4
		5	8	457	57.1	59.4			
	406	4	5	1,750	350.0	234.5	4,800	960.0	709.2
		5	7	1,305	186.4	166.2			
	407	4	0						
		5	2	360	180.0	169.7			
	408	4	7	5,940	848.6	988.9	5,130	732.8	565.9
		5	12	2,025	168.8	159.2			
	444	4	14	4,650	332.1	635.7	8,590	613.6	644.6
		5	12	550	45.8	72.2			
	478	4	12	2,200	183.3	156.3	5,200	433.3	376.8
		5	12	1,190	99.2	60.2			
	544	4	11	12,600	1,145.4	2,045.5	8,850	804.5	776.7
		5	16	2,442	152.6	233.9			
	550	4	1	100	n.a.	n.a.	250	n.a.	n.a.
		5	6	1,135	189.2	299.8			
	562	4	1	250	n.a.	n.a.	300	n.a.	n.a.
		5	8	1,685	210.6	268.6			
	571	4	6	740	123.3	75.6	2,048	341.3	218.1
		5	6	1,110	185.0	39.4			
		4	10	3,083	308.3	187.5	5,715	571.5	243.0
		5	18	2,107	117.0	98.9			
Habersham		4	3	175	58.3	80.4	650	216.7	152.8
		5[a]	48	2,315	48.2	37.6			

County	ID							
Harris								
	672	4	24	153,101	6,379.2		36,465	1,519.4	
		5	75	75,608	1,008.1				
		4	1	7,000	n.a.		1,200	n.a.	n.a.
	679	4	3	250	83.3				
		5	4	1,800	450.0		2,500	625.0	394.8
		4	11	2,535	230.4				
	695	5	4	86,135	21,533.8		10,065	2,516.2	1,638.4
		4	5	14,010	2,802.0	15,313.6			
		5	2	1,000	500.0	3,419.7			
	696	4	4	1,075	268.8	0.0	1,400	700.0	141.4
		5	6	10,900	1,816.7	302.3			
	703	4	19	33,310	1,753.2	3,040.0	8,000	1,333.3	781.4
		5	1	14,380	n.a.	3,139.3			
	707	4	3	150	50.0	n.a.	3,000	n.a.	n.a.
		5	1	200	n.a.	0.0			
	717	4	6	528	88.0	n.a.	400	n.a.	n.a.
		5	0			80.2			
	781	4	12	10,100	841.7	1,791.5			
		5	0						
	782	4	0						
		5	1	0	n.a.	n.a.	4,300	n.a.	n.a.
	786	4	4	650	162.5	103.1			
		5	0						
	920	4	2	5,300	2,650.0	3,323.4			
		5	0						
	934	4	0						
		5	4	31,686	7,921.5	13,057.4	5,600	1,400.0	867.9
	1186	4	6	7,700	1,283.3	3,070.1			
Houston		4	12	33,693	2,807.8	7,959.2	17,400	1,450.0	3,334.3
		5	59	30,948	524.5	1,332.7			
Thomas		4	77	734,970	9,545.1	38,680.3	246,095	3,196.0	7,108.1
		5	148	97,984	662.0	2,353.8			

(*continued*)

a Contains two records with "tenant" in the occupation column on schedule I and no real property. These two records do not appear on schedule IV.

Appendix H (Continued)

County	Militia District	Type	No.	Total Personal Property ($)	Mean Personal Property ($)	Stand. Dev. Personal Property ($)	Total Real Property ($)	Mean Real Property ($)	Stand. Dev. Real Property ($)
Towns		4	1	400	n.a.	n.a.	900	n.a.	n.a.
		5	6	340	56.7	71.2			
Twiggs		4	0						
	323	5	23	945	41.1				
	324	4	0						
		5	0						
	325	4	0						
		5	2	150	75.0	35.4			
	326	4	0						
		5	4	0	n.a.	n.a.			
	354	4	0						
		5	0						
	355	4	0						
		5	4	150	37.5	25.0			
	356	4	0						
		5	6	520	86.7	84.4			
	372	4	0						
		5	4	125	31.2	23.9			
	396	4	0						
		5	1	0	n.a.	n.a.			
	425	4	0						
		5	2	0	n.a.	n.a.			
Wilkes		4	23	572,990	24,912.6	32,357.4	310,412	13,496.2	28,819.3
		5	19	61,050	3,213.2	4,165.5			
Worth		4	7	9,880	1,411.4	2,156.1	5,003	714.7	637.9
		5	6	1,656	276.0	344.6			

Notes

CHAPTER 1: INTRODUCTION

1. Two fundamentally opposed conceptions of slavery and antebellum southern society are represented by Eugene D. Genovese, *The Political Economy of Slavery: Studies in the Economy and Society of the Slave South* (New York, 1965); and Robert William Fogel and Stanley L. Engerman, *Time on the Cross: The Economics of American Negro Slavery* (Boston and Toronto, 1974). The classic statement of postbellum discontinuity remains C. Vann Woodward, *Origins of the New South, 1877–1913* (Baton Rouge, La., 1951); but for one recent forceful challenge, see Jonathan M. Wiener, *Social Origins of the New South: Alabama, 1860–1885* (Baton Rouge, La., 1978). An important study that covers the entire Civil War era and emphasizes discontinuity is Steven Hahn, *The Roots of Southern Populism: Yeoman Farmers and the Transformation of the Georgia Upcountry, 1850–1890* (New York and Oxford, 1983). A collection of essays with ample bibliography is Walter Fraser and Winifred B. Moore, eds., *From the Old South to the New: Essays in the Transitional South* (Westport, Conn., 1981).

2. Hahn, *Roots of Southern Populism*, 161–65; Jonathan M. Wiener, "Class Structure and Economic Development in the American South, 1865–1955," *American Historical Review* 84 (Oct. 1979): 992; and "Comment" by Harold D. Woodman, ibid., 997–1001. The most important studies by economists who judge tenancy positively are Stephen J. DeCanio, *Agriculture in the Postbellum South: The Economics of Production and Supply* (Cambridge, Mass., 1974); Robert Higgs, *Competition and Coercion: Blacks in the American Economy, 1865–1914* (Cambridge, Eng., 1977); and several articles by Joseph D. Reid, Jr., especially "White Land, Black Labor, and Agricultural Stagnation: The Causes and Effects of Sharecropping in the Postbellum South," *Explorations in Economic History* 16 (Jan. 1979): 31–55. Economists who argue, however, that racism significantly distorted the market are Roger L. Ransom and Richard Sutch, *One Kind of Freedom: The Economic Consequences of Emancipation* (Cambridge, Eng., 1977). See also the critique of the economists by Harold D. Woodman, "Sequel to Slavery: the New History Views the Postbellum South," *Journal of Southern History* 43 (Nov. 1977): 523–54.

3. Ransom and Sutch, *One Kind of Freedom*, 88; Wiener, *Social Origins of the New South*, 93; Hahn, *Roots of Southern Populism*, 22–23; and Joseph D. Reid, Jr., "Antebellum

Southern Rental Contracts," *Explorations in Economic History* 13 (Jan. 1976): 69–83. For a good review of recent studies on landholding in the United States, see Robert P. Swierenga, "Quantitative Methods in Rural Landholding," *Journal of Interdisciplinary History* 13 (Spring 1983): 787–808.

4. Frank L. Owsley, *Plain Folk of the Old South* (Chicago, 1965 [first published, 1949]), 16; Ransom and Sutch, *One Kind of Freedom*, 88, 336 n. 19; and Gavin Wright, *The Political Economy of the Cotton South: Households, Markets, and Wealth in the Nineteenth Century* (New York, 1978), 39.

5. A more detailed discussion of the Owsley school's method can be found in chapter 2.

6. Morton Rothstein, "The Antebellum South as a Dual Economy: A Tentative Hypothesis," *Agricultural History* 41 (Oct. 1967): 373–83; Eugene D. Genovese, "Yeoman Farmers in a Slaveholders' Democracy," *Agricultural History* 49 (April 1975): 331–42; Hahn, *Roots of Southern Populism*.

7. For an extensive treatment of westward movement using the manuscript census and based on the Parker-Gallman Sample, see James D. Foust, "The Yeoman Farmer and the Westward Expansion of U.S. Cotton Production," Ph.D. dissertation, University of North Carolina, 1967 (reprinted by Arno Press, 1976).

8. See, for example, Richard A. Easterlin et al., "Farms and Farm Families in Old and New Areas: The Northern States in 1860," in *Family and Population in Nineteenth-Century America*, Tamara K. Hareven and Maris A. Vinovskis, eds. (Princeton, N.J., 1978), 22–84.

9. For a fuller account of these and other findings, see Frederick A. Bode and Donald E. Ginter, "A Critique of Landholding Variables in the 1860 Census and the Parker-Gallman Sample," *Journal of Interdisciplinary History* 15 (Autumn 1984): 277–95. For a recent review of southern data bases, including the Parker-Gallman Sample, and the literature that has emerged from them, see Swierenga, "Quantitative Methods in Rural Landholding," 791–95.

10. Donghyu Yang, "Notes on the Wealth Distribution of Farm Households in the United States, 1860: A New Look at Two Manuscript Census Samples," *Explorations in Economic History* 21 (Jan. 1984): 88–102.

11. Enoch Marvin Banks, *The Economics of Land Tenure in Georgia* (New York, 1905), 82–83.

12. Marjorie Stratford Mendenhall, "The Rise of Southern Tenancy," *Yale Review*, n.s. 27 (Sept. 1937): 116–17. Mendenhall allowed some importance to antebellum tenancy although Ransom and Sutch correctly point out that she presents her case "rather unconvincingly." *One Kind of Freedom*, 336 n. 20.

13. Lewis Cecil Gray reached his conclusions on the basis of a few newspaper advertisements but was more impressed by the "scarcity of references" to tenancy. *History of Agriculture in the Southern United States to 1860* (Gloucester, Mass., 1958 [first published, 1933]), vol. 2, 646–47. State studies of antebellum southern agriculture virtually ignore tenancy. A footnote reference to it can be found in Charles S. Davis, *The Cotton Kingdom in Alabama* (Philadelphia, 1974 [first published, 1939]), 42 n. 145. A brief discussion of tenancy can also be found in Guion Griffis Johnson, *Ante-Bellum North Carolina: A Social History* (Chapel Hill, N.C., 1937), 68–69.

14. Reid, "Antebellum Southern Rental Contracts." For evidence from antebellum

North Carolina court records on renting and sharecropping, see Marjorie Mendenhall Applewhite, "Sharecroppers and Tenants in the Courts of North Carolina," *North Carolina Historical Review* 31 (April 1954): 134–49.

15. Harold D. Woodman, "Post-Civil War Southern Agriculture and the Law," *Agricultural History* 53 (Jan. 1979): 325–26.

16. Ransom and Sutch, *One Kind of Freedom*, 88.

17. Gavin Wright, "The Strange Career of the New Southern Economic History," *Reviews in American History* 10 (Dec. 1982): 165.

CHAPTER 2: METHOD

1. The problem of defining the various forms of tenancy and the adequacy of the 1880 census classification is discussed in chapter 5.

2. These other schedules are schedule II, Slave Inhabitants; schedule III, Statistics of Mortality; schedule V, Productions of Industry; and schedule VI, Social Statistics. The last schedule includes aggregate data for each county on taxation, education, religion, newspapers, pauperism, crime, and wages. In addition to schedules I and IV, the most relevant for the study of tenancy is schedule II, which reveals the extent of slaveholding (though not necessarily slave ownership) among tenants. A consideration of slaveholding in relation to tenancy is not included in the present study. The censuses of 1850 and 1870 are similar in most respects to that of 1860, and most of the comments here apply to them as well. Obviously, there was no slave schedule in 1870. The standard source for information on the census through 1890 is Carroll D. Wright (assisted by William C. Hunt), *The History and Growth of the United States Census, Prepared for the Senate Committee on the Census* (Washington, D.C., 1900). The 1860 schedules are reproduced in this volume, but not the instructions to the enumerators, because no copies of the latter could be found at the time (ibid., 131). However, Lee Soltow in his *Patterns of Wealthholding in Wisconsin since 1850* (Madison, Wis., 1971), 156 n. 2, has pointed out that a copy of the instructions, of which we have used a photocopy, is available in the library of the American Philosophical Society in Philadelphia (U.S. Census Office, Eighth Census [1860], *Instructions to U.S. Marshalls. Instructions to Assistants* [Washington, D.C., 1860]). All subsequent references to the 1860 census instructions are to this source.

3. A discussion of the relevant literature can be found below.

4. The general problem of the reliability and usefulness of individual census variables is discussed in chapter 3.

5. The problem of classifying operators according to production is discussed below.

6. James D. Foust and Dale E. Swan incorrectly assert that farmers' sons are not designated on the census as "farm laborers." They are often so designated, the principal alternative being simply "farmer." Enumerators are normally consistent in this respect within their county. Some enumerators also listed all farmers' wives as "farmer," while others listed them as "keeping house" or gave them no occupation at all. See Foust and Swan, "Productivity and Profitability of Antebellum Slave Labor: A Micro-Approach," *Agricultural History* 44 (Jan. 1970): 41. The inconsistency of enumerators in reporting occupations renders the state-level occupational breakdowns in the published census useless for many purposes.

7. The remuneration of assistant marshals at ten cents for each farm enumerated, as opposed to two cents for each person, possibly accounts for their overzealousness.

8. In general when we refer, for example, to the percentage of certain tenants (type 2) in a county, we include only the number of type 2a entries in the numerator, and the total of all schedule IV entries less *all* types b and c on the schedule in the denominator. In this way we eliminate the bias resulting from a disproportionate number of types b and c in any particular category of operator. An estimation of the incidence of types b and c may also be inferred from appendix A, where short-method estimates of all tenants and of type 2a only are presented for all counties of Georgia.

9. Imposing arbitrary cutoffs (that is, without using price data) on arable production would have been almost impossibly complex in any case. It would have necessitated establishing an upper limit for each crop, if only one crop were produced, and progressively lower limits if two, three, or more crops were produced.

10. Seddie Cogswell, Jr., *Tenure, Nativity and Age as Factors in Iowa Agriculture, 1850–1880* (Ames, Iowa, 1975), 8–9.

11. In addition to applying different standards with respect to production, enumerators were inconsistent in applying the census instructions on the status of the farm operator with respect to ownership, management, and tenancy. See chapter 5.

12. If an enumerator included all first-year farmers on schedule IV, regardless of the level of their production, almost certainly some of them, as in Spalding, would have appeared as type b or c; yet, as already noted, eight of our sample counties contain no such entries. Some counties we have examined have virtually no operators without arable production. On the other hand, counties with small numbers of types b and c may be reporting first-year farmers but not town dwellers or overseers who possessed some livestock.

13. Spalding is located in the western lower Piedmont, and how typical a county it was in 1860 is problematic. Because it was created in 1851, we cannot compute the rate of change of any variable between 1850 and 1860. It is perhaps suggestive that the population of Butts, just to the east, remained practically stationary between the two census years, while that of Coweta, just to the west, increased only by just under 8 percent. (The boundaries of these two counties were not altered between 1850 and 1860.) The population of the state as a whole, however, grew by almost 17 percent. One might expect, therefore, that in the more rapidly growing parts of the state, such as the southwestern cotton belt, the proportion of first-year farmers would be greater than in Spalding.

14. This confusion was particularly apparent in some Georgia counties where the enumerator failed to enter tenants on schedule IV at all. See chapter 2.

15. The possibility that farm operators without real property were agents or managers has been suggested by Herbert Weaver, *Mississippi Farmers, 1850–1860* (Gloucester, Mass., 1968 [first published, 1945]), 64; and Randolph B. Campbell, "Planters and Plainfolk: Harrison County, Texas, as a Test Case, 1850–1860," *Journal of Southern History* 40 (Aug. 1974): 369–98.

16. Table 2.2 shows that in a few counties manager or agent appears very infrequently on schedule IV. It is possible that these counties had only one or two managed farms. But this seems unlikely in a county such as Clay, located in the newer cotton lands of the southwest, where one would expect some absentee ownership. Similarly,

the presence of only one designated tenant in Habersham and three designated tenants in Gwinnett is suspect, to say the least. We are certain that anyone who has had a close experience with these returns will agree that some enumerators were highly erratic in one or more (and seldom the same) respect.

17. In Gwinnett County there was considerable variation among districts in the reporting of acreage. In other counties there were only a few exceptions to the dominant conventions shown in table 2.2.

18. A more detailed discussion of tenancy conventions and their possible meaning follows in this chapter and in chapter 5.

19. We have disregarded the presence or absence of unimproved acreage in assigning entries in our sample counties to the type 3 category. We have confirmed that missing unimproved constitutes a distinct tenancy convention in some counties examined by the short method, but since some tenants, like some owner-operators, may simply not have held unimproved acres, the convention should not generally be applied.

20. The phrase "'farmer' without a farm" was first used by Merle Curti, *The Making of an American Community: A Case Study of Democracy in a Frontier Community* (Stanford, Calif., 1959), 59–60. We, like Curti, believe that the phrase is an appropriate description of schedule I "farmers" who do not appear on schedule IV (types 4 and 5), and have so used it in this study. Donald L. Winters has applied the term to tenants on schedule IV who were, however, clearly farmers *with* farms, even if they did not own them. See his study *Farmers without Farms: Agricultural Tenancy in Nineteenth Century Iowa* (Westport, Conn., and London, 1978).

21. We have also included in the type 4 and type 5 categories farmers appearing on schedule I only who gave a second occupation, such as "farmer and lawyer."

22. In Gwinnett County the mode of reporting tenants, and therefore our application of the rules, varied among districts. Differences in handwriting, as well as unusually sharp variations in procedure, suggested different enumerators. It is also possible that in some counties type 2 records, defined by our rules as defective, may not be defective at all. They may represent alternative tenancy conventions.

23. We did correct some obvious errors. For example, enumerators occasionally skipped too many, or not enough, lines when entering data for a farm operator on the second page of schedule IV. This resulted in alignment problems between the two pages. When this occurred only to a minor degree, the correct alignment was usually obvious. Since we did not use any variables from the second page in our analysis, errors we might have made in correcting the alignment would only very slightly affect our count of types b and c; however, researchers using variables on the second page should delete any records for which they have serious doubts about the alignment. It was also usually obvious whenever a particular variable was *consistently* recorded in the wrong column on a page.

24. In the short method we counted designated managers and agents as owner-operators. Obviously, we could not identify type 6 records—schedule IV entries not traceable to schedule I.

25. See chapter 5.

26. The approximation of types 3 and 5 is crude because they could be over- or underrepresented on any page of the schedule. Also, since some names on one schedule are sometimes out of sequence on the other schedule, even when the schedules are generally in phase, we might simply have failed to find the match.

27. Since we did not make a full examination of all schedule I pages for most counties, the information on tenants designated on that schedule may be incomplete.

28. In a small number of counties variation in conventions is sectored and seems to have resulted from the fact that more than one enumerator was responsible for the returns. In most counties the conventions are distributed throughout the return, even when multiple handwritings are present. They are also normally present on the same page, suggesting they are not a function of local variation in reporting modes but a reflection of real tenancy differences. For a fuller discussion of these points, see chapter 5.

29. We have identified missing unimproved acreage as a tenancy convention in some counties; but in those counties there are usually also a small number of owner-operators who have no unimproved acreage and are thus mistakenly counted as tenants by the short method. Missing unimproved acreage is in fact the most unreliable of the tenancy conventions.

30. The results of all the studies can be found in Frank L. Owsley, *Plain Folk of the Old South* (Chicago, 1965 [first published, 1949]). See also Frank L. Owsley and Harriet C. Owsley, "The Economic Basis of Society in the Late Ante-Bellum South," *Journal of Southern History* 6 (Feb. 1940): 24–45, and "The Economic Structure of Rural Tennessee, 1850–1860," ibid. 8 (May 1942): 161–82; Blanche Henry Clark, *The Tennessee Yeomen, 1840–1860* (New York, 1971 [first published, 1942]); Harry L. Coles, Jr., "Some Notes on Slave Ownership and Land Ownership in Louisiana, 1850–1860," *Journal of Southern History* 9 (Aug. 1943): 381–94; and Weaver, *Mississippi Farmers*. The leading critics of the Owsley studies are Fabian Linden, "Economic Democracy in the Slave South: An Appraisal of Some Recent Views," *Journal of Negro History* 31 (April 1946): 140–89; and Gavin Wright, "'Economic Democracy' and the Concentration of Agricultural Wealth in the Cotton South, 1850–1860," *Agricultural History* 44 (Jan. 1970): 63–85. Owsley's summary estimate of the rate of landownership, 80 to 85 percent (*Plain Folk*, 16), has been accepted by Roger L. Ransom and Richard Sutch, *One Kind of Freedom: The Economic Consequences of Emancipation* (Cambridge, Eng., 1977), 88, 336 n. 19; and Gavin Wright, *The Political Economy of the Cotton South: Households, Markets, and Wealth in the Nineteenth Century* (New York, 1978), 39. Actually, Owsley's figure is not consistent with his own data, which suggest a much lower rate, more like 70 to 75 percent. We have calculated rates of landownership for the eighteen regions studied by the Owsley group for 1850 and 1860, using their data. See appendix D.

31. A more detailed comparison of Owsley's Georgia results with our own can be found in chapter 6.

32. Two recent studies of tenancy in Iowa, for example, exclude types 4 and 5 in calculating tenancy rates. See below.

33. Owsley noted (*Plain Folk*, 193 n. 1) that Weaver's classification differed "somewhat" from Clark's and his own.

34. Weaver, *Mississippi Farmers*, 15–16.

35. Such exclusion of type 4 entries would have been consistent with Weaver's assumptions. He quite plausibly believed that many such entries were persons who resided in one county but owned their farm in another county. Since he assumed that enumerators typically entered only landowners on schedule IV, to have included type 4

in his computations might have resulted in double counting, that is, counting a type 4 once from schedule I in his county of residence and a second time from schedule IV in the county where he owned a farm.

36. See table 2.1 for the proportion of types b and c in the sample Georgia counties, and appendix A for the differences between type 2a and all type 2 entries in the Georgia counties examined by the short method.

37. Another pioneering study using the manuscript census to examine landholding patterns in Hancock County, Georgia, in 1850 and 1860, which appeared about the same time that the Owsley group was publishing its findings, is James C. Bonner, "Profile of a Late Ante-Bellum Community," *American Historical Review* 49 (July 1944): 663–80. Bonner, unfortunately, never clarified his method. He noted that in 1850, 210 "farmers," presumably on schedule I only, owned no real estate (out of how many farmers altogether?) and that the enumerator "occasionally" designated renters. He also pointed out that "many" landless farmers appeared on schedule I in 1860, none of whom appeared on schedule IV. Bonner concluded that these findings indicated a shift from a reliance on tenants to a reliance on farm laborers who were identified as farmers on schedule I. It is just as plausible that the 1860 enumerator failed to enter tenants on schedule IV so that they appear as type 5 farmers that year.

38. Curti, *The Making of an American Community,* 59–60.

39. Ibid., 144.

40. Allan G. Bogue, *From Prairie to Cornbelt: Farming on the Illinois and Iowa Prairies in the Nineteenth Century* (Chicago, 1963), 64. Bogue did not mention the presence of type 6, type c, or any equivalent to our type b.

41. This discussion is based on the methodological comments of Cogswell, *Tenure, Nativity and Age,* 6–13; and Winters, *Farmers Without Farms,* 12–13 and 109–13. Cogswell examined six contiguous counties for the period from 1850 to 1880 but after 1850 confined himself to twenty-six contiguous townships in those counties. For the same period Winters employed "a stratified, subsampling procedure" to choose twelve counties out of each of which, after 1850, he randomly selected five townships.

42. See our discussion of Cogwell's inference above.

43. According to Cogswell, most types 4 and 5 were probably first year farmers with insufficient production to qualify for schedule IV. He maintained that they were less frequently found in counties with our types b and c operators. As we shall later argue in chapter V, this pattern does support his conclusion that types 4 and 5 were first year farmers.

44. In chapter 5 we will discuss the status of farmers without farms at greater length and will consider the alternatives of including or excluding them in a tenancy count.

45. Randolph B. Campbell and Richard G. Lowe, *Wealth and Power in Antebellum Texas* (College Station, Texas, and London, 1977), especially 13–31.

46. Ibid., 67–68. Campbell and Lowe explicitly state (21) that farmers "without farms" are schedule I "farmers" not traceable to schedule IV. The percentages in the quote are based on tables (28 and 67) using the designation "without farms."

47. Campbell and Lowe do note (ibid., 20) that some "enumerators refused to record the cash value of the farm if the farm operator was not the owner." At no point do they mention the presence of farm operators with only minor livestock or no produc-

tion (our types b and c). An earlier study of Harrison County, Texas, by Campbell ("Planters and Plainfolk") exhibits many of the same methodological problems as the larger eastern Texas study.

48. Frank Jackson Huffman, Jr., "Old South, New South: Continuity and Change in a Georgia County, 1850–1880" (unpublished Ph.D. dissertation, Yale University, 1974), 271–73.

49. Ibid., 69–70. It may be appropriate for certain purposes to include all persons with a given occupation in a single category, but the lack of a separate category for heads of household prohibits a comparison of Huffman's results with those of the great majority of studies that consider heads of household only.

50. Ibid., 273.

51. Steven Hahn, *The Roots of Southern Populism: Yeomen Farmers and the Transformation of the Georgia Upcountry, 1850–1890* (New York and Oxford, 1983), 21–23, 291–95.

52. Hahn's argument for informality also rests on the presence of alleged "kin tenants," landless farmers listed on the census next to or near a landholding head of household with the same surname. Such tenants accounted for about 25 percent of landless farmers in both Carroll and Jackson counties before the Civil War (p. 23). Quite apart from the inherent unreliability of such an estimate, Hahn's case for a fundamental change in the nature of tenancy after the Civil War is not helped by the fact that he cites the same figure—one quarter—for the proportion of kin tenants in 1880 (p. 165).

CHAPTER 3: THE CENSUS VARIABLES: A CRITIQUE

1. We did not include these variables for the balance of Georgia's counties examined by the short method, since the purpose of that method was only to obtain an estimate of tenancy rates.

2. For the sake of consistency and clarity we did make some alteration in the designation of occupations, but only when we were certain that we were not obscuring distinctions intended by the enumerator. For example, a single enumerator occasionally recorded "farms" or "farming" as an occupation, as well as "farmer." We changed such cases to the more common designation, farmer. In this example and in other such cases that we standardized we could find no evidence in the returns that the enumerator was being anything but inconsistent in recording the information. For additional discussion of implausible values for variables, see chapter 2.

3. There is an extensive literature on the "agricultural ladder," a metaphor for a presumed mobility sequence extending from farm laborers on the bottom rung to proprietors of large holdings on the top rung. A useful review of this literature can be found in Donald L. Winters, *Farmers Without Farms: Agricultural Tenancy in Nineteenth Century Iowa* (Westport, Conn., and London, 1978), 78–91. A full consideration of mobility, which we have not attempted, would require time-series data on farm laborers, tenants, and owner-operators for at least two census years. Alternatively, as a first approximation, one could employ age data, which we have not gathered for the present study.

4. We have developed tables showing the occupational breakdown for each identified type number at the county level or, where available, at the militia-district level in the sixteen sample counties. Copies of these tables can be obtained from the authors.

5. All references to the census instructions are to U.S. Census Office, Eighth Census [1860], *Instructions to U.S. Marshalls. Instructions to Assistants* (Washington, D.C., 1860), photocopy obtained from the library of the American Philosophical Society, Philadelphia.

6. In all counties we examined there were some farm operators who were given a nonagricultural occupation, such as lawyer or merchant. This would not necessarily have been a violation of the instructions so long as a person, who was, for example, a merchant in his principal occupation, also operated a farm. In Baldwin County four individuals were called planter on schedule I but reported no real property and were explicitly identified as renters on schedule IV.

7. Most enumerators did distinguish "laborer" (or "common laborer," "day laborer," or "hireling") from "farm laborer," although a few enumerators designated farm laborer as "farm hand," technically a violation of the instructions. It may be that in rural communities there was little real distinction between farm labor and common labor. Some persons may have drifted between town and countryside in search of casual employment; however, occupational distinctions of this kind in a single county and even in a single household suggest that some enumerators intended to point to real differences. But while distinctions within counties may be real, distinctions among counties resulting from enumerator inconsistency should not be relied upon.

8. The only nonheads of household in our data base are the very infrequent schedule IV operators who did not head a household. See also chapter 2.

9. There were some tenants, as there were some owner-operators, who reported nonagricultural occupations.

10. In a few counties there were only one or two cases of such designations, and they could be errors. In a very few instances even an owner-operator with production (type 1a) was placed in a laboring occupation.

11. Tenancy designations can also appear in the real property column of schedule I or on schedule IV. Since we have not made a complete search of schedule I for most Georgia counties, there may be others besides the ones mentioned that include tenancy designations in the occupation column.

12. It will be recalled that many, perhaps most, farm laborers in some counties were not heads of household but were apparent boarders in the households of farm operators.

13. Type 5—and type 4—records could also have been first-year farmers with no production to report on schedule IV. For a full discussion of these problems, see chapter 5.

14. This rule is subject to restrictions regarding explicitly managed farms and schedule IV entries qualifying as possible tenants (type 3). The value of real property is also used to distinguish type 4 from type 5. See chapter 2.

15. See table 2.2. For the reporting of farm value as part of a tenancy convention in counties examined by the short method, see appendix A.

16. Ulrich Bonnell Phillips, *American Negro Slavery: A Survey of the Supply, Employment and Control of Negro Labor as Determined by the Plantation Regime* (Baton Rouge, La., 1966 [first published, 1918]), 288. See also Kenneth M. Stampp, *The Peculiar Institution: Slavery in the Antebellum South* (New York, 1956), 43–44.

17. All but one of the thirty-five type 6 entries in Baker were reported in two con-

tiguous blocks on schedule IV at the end of each postal district. The great wealth of all but a few of these operators makes it unlikely that they are simply missing from schedule I. Also, some of the names can be identified as persons residing elsewhere. In other counties it was not characteristic for type 6 entries to appear in blocks.

18. A few enumerators in Georgia did enter the names of joint owners on schedule IV.

19. Possible enumerator errors include leaving off a digit for one of the two variables. In Houston County, for example, one farm operator reported $29,000 real and $2,900 farm value. Another possible type of error involved misplacing digits in one of the variables. Thus an enumerator recorded $7,930 real and $7,093 farm value for a farmer in Greene County. Sometimes the values of the two variables differed by only a very small amount, such as $21,710 real and $21,715 farm value entered for a farmer in Twiggs County. In one case in Gwinnett County the two values differed by only one dollar. There were also a few cases where farm value equalled personal property but not real property. Every county contained a number of possible errors of these kinds. We have not attempted to correct any such apparent errors in our data, and they are reflected in table 3.2.

20. Georgia, Office of the Comptroller General, *Annual Report of the Comptroller General, of the State of Georgia, Made to the Governor, October 20, 1860* (Milledgeville, 1860), table A, 38. Hereafter these reports are cited as *Comptroller General Report* with the relevant year.

21. The validity of these calculations depends on the assumption that the census and the comptroller general were counting roughly the same acreage. But in the state as a whole and in many counties this was clearly not the case. The number of taxable acres in the state in 1860, according to the comptroller general, was 33,345,289, while the number of farm acres, according to the census, was 26,650,490. This difference is not implausible if we assume that the number of taxable acres was much greater than the number of farm acres; but more disconcerting as far as the reliability of the comptroller general's data is concerned, is the fact that the number of taxable acres in many counties fluctuated wildly from year to year. Between 1859 and 1860, when no new counties were created, there was a difference of less than 5 percent (plus or minus) in the taxable acreage reported in only 69 of Georgia's 132 counties. In 32 counties the difference was 10 percent or greater, in 15 counties 20 percent or greater, and in 7 counties 30 percent or greater. Of these 32 counties, 13 showed a *loss* of taxable acreage of greater than 10 percent. In Emanuel County, however, the reporting was at least consistent. Taxable acreage remained practically unchanged between 1857 and 1858. In 1858 Emanuel lost territory to new counties, a loss reflected in a drop in its taxable acreage in 1859. But again the taxable acreage varied only slightly, increasing by 2.8 percent, between 1859 and 1860.

22. It would have been unlikely indeed for an enumerator who had to collate at least two and frequently three schedules for hundreds of farm operators to make no errors. An error ratio of 3.8 percent with respect to real-property and farm value for 559 type 1 farm operators in Harris County does not seem extraordinarily high. These possible errors include some examples of the type discussed in note 19, above.

23. Muscogee County also borders on Harris, but since it lies largely in a different soil region, the Sand Hills, and contains what was already by 1860 the sizable town of Columbus, it is excluded from the table.

24. This generalization does not apply to improved acres as a percentage of total farm acres. This variable is included on the table only to determine if a disproportionate number of improved acres in any county might have driven farm value per total acres upward. It is clear that such a bias is not present.

25. The various regions of Georgia are discussed more fully in chapter 4.

26. For examples of the traditional view of antebellum tenancy, see Enoch Marvin Banks, *The Economics of Land Tenure in Georgia* (New York, 1905), 8–9; Guion Griffis Johnson, *Ante-Bellum North Carolina: A Social History* (Chapel Hill, N.C., 1937), 68–69; Marjorie Stratford Mendenhall, "The Rise of Southern Tenancy," *Yale Review* n. s. 27 (Sept. 1937): 110–29; and Willard Range, *A Century of Georgia Agriculture, 1850–1950* (Athens, Ga., 1954), 7. See also Roger L. Ransom and Richard Sutch, *One Kind of Freedom: The Economic Consequences of Emancipation* (Cambridge, Eng., 1977), 88. Ransom and Sutch (336 n. 21) have also discounted Joseph D. Reid, Jr.'s examination of tenancy contracts from an upland county in western North Carolina as being inapplicable to the cotton South. Reid, "Antebellum Southern Rental Contracts," *Explorations in Economic History* 13 (Jan. 1976): 69–83. Frank L. Owsley, while not suggesting that tenants were, in effect, planters, did reject the poor white hypothesis. *Plain Folk of the Old South* (Chicago, 1965 [first published, 1949]), 8.

27. Beginning in 1803, public lands in Georgia, comprising most of the state at that time, were distributed by a series of lotteries. Not all such lands were taken up, however. In 1857 the comptroller general complained that a "considerable amount" of land that was not returned for taxes probably belonged to persons not residing in the state or was lottery land that "may never [have been] claimed by their proper owners." *Comptroller General Report* (1857), 18–19. The best overview of the Georgia land system can be found in Milton Sydney Heath, *Constructive Liberalism: The Role of the State in Economic Development in Georgia to 1860* (Cambridge, Mass., 1954), 69–92, 139–58. It is also quite likely that some nonresidents of a county or even urban residents of a county had invested in lands that they had not yet begun to clear and farm or were holding for speculative purposes. Unless these persons actually operated a farm in the county, such acreage as they owned would presumably not have appeared on schedule IV.

28. See earlier discussion in this chapter.

29. A low ratio could reflect, in addition to the type of error discussed here, either the extensive cultivation of cotton at the expense of other crops and grazing land for livestock, or its intensive cultivation on good soils and with optimum management practices. The 1880 census was the first to include the number of acres planted in each crop.

30. The cotton-intensity ratios on the two parts of table 3.5 are not strictly comparable, because the data for the lower part is taken from the aggregate census and thus includes the improved acreage of all operators, not just cotton producers. Hancock, Jones, Putnam, Washington, and Wilkinson counties border on Baldwin, the first three of which are in the Piedmont and the last two, in the Central Cotton Belt. Bibb and Glascock are nearby Sand Hills counties to the west and east, respectively, of Baldwin.

31. If one takes the percentage of improved farm acres for the five counties bordering Baldwin (43.6 percent) and adjusts Baldwin's improved acres upward accordingly, the result is a cotton-intensity ratio for Baldwin of 10.2, which is much more in line with the region: $(.436 \times 159{,}826$ total farm acres in Baldwin$)/(6{,}811$ cotton bales in Baldwin$) = 10.2$.

32. For a discussion of schedule IV tenancy conventions, see chapter 2.

258   Notes

33. See table 2.2. Wilkes did not report farm value for tenants, whereas Worth regularly reported all three variables and explicitly designated tenants as such on schedule IV. Class 4 counties, examined by the short method and shown in appendix A, also apparently reported full values for tenants.

34. R. H. Clark, T. R. R. Cobb, and D. Irwin, *The Code of the State of Georgia* (Atlanta, 1861), 150–51. It is unlikely that many tenants would have been the legal agent of their landlord.

35. This may have been strictly true only for rented farms as opposed to farms that were operated on sharecropping arrangements and may explain why some enumerators might have excluded the latter from schedule IV altogether. See chapter 5.

36. Emphasis added in this sentence and in other quotations from the census instructions in this paragraph.

37. Even though the estimated 4,300 tenanted acres exceeds the total acreage for proprietors of 200 or more acres, we assume that only some unknown, if substantial, proportion of tenanted acreage belonged to these proprietors. Some tenanted acreage almost certainly belonged to smaller proprietors. Moreover, an estimate of 25 improved acres for mean tenanted farm size in Towns is probably too high. In the upper Piedmont county of Gwinnett, improved acreage was reported for tenants in four militia districts. In those districts mean tenanted farm size (type 2) was 25.4 improved acres, while owner-operators (type 1) had a mean of 71.6 improved acres, substantially higher than the mean of 56.2 for type 1 operators in Towns. Improved acres for type 1 operators in Towns, unlike in the four Gwinnett districts, in all likelihood also included tenanted acreage as well. A more realistic mean farm size for tenants in Towns was probably about 20 improved acres.

38. Mean tenanted farm size probably fell somewhere between 30 and 50 improved acres in Twiggs. Mean farm size for type 2 tenants was, in those cotton belt counties that reported it, 43.4 improved acres in Baker, 31.5 in Baldwin, and 86.0 (only thirteen cases) in Wilkes. If tenants averaged 40 improved acres in Twiggs, and these acres constituted the same proportion of total acreage (improved plus unimproved) as they did for proprietors (44.4 percent), then tenants would have averaged about 90 total farm acres.

39. If the enumerator did make such entries, we would have categorized them as types 1c and 6c. But most counties in our sample have no operators at all in these categories, and when they do, they are few, with only small amounts of acreage.

40. Robert William Fogel and Stanley L. Engerman, *Time on the Cross: Evidence and Methods—A Supplement* (Boston and Toronto, 1974), 126–51. For a critique of Fogel and Engerman's conceptualization of the efficiency problem, see Paul A. David et al., *Reckoning with Slavery: A Critical Study in the Quantitative History of American Negro Slavery* (New York, 1976), 202–23.

41. In other words, decreasing the ratio of labor and capital to land, while holding output constant, will have the effect of increasing the productivity of the first two factors.

42. The question of whether individuals entered on schedule II as slaveholders included both owners and hirers, or just owners, in any particular county is exceedingly complex. We will treat this matter at greater length in a forthcoming study.

43. Our data show that large numbers of wealthy planters who owned much of their

land outside of Baldwin lived in the state capital, Milledgeville, or the resort town of Scottsboro.

44. It should be noted that the possible inconsistencies in personal property between Greene and Wilkes discussed above will have little if any impact on the values reported in the lower ranges on table 3.8.

45. *The Political Economy of the Cotton South: Households, Markets, and Wealth in the Nineteenth Century* (New York, 1978), 43–88. On the problem of antebellum self-sufficiency in foodstuffs, see Robert E. Gallman, "Self-Sufficiency in the Cotton Economy of the Antebellum South," *Agricultural History* 44 (Jan. 1970): 5–23; and Sam Bowers Hilliard, *Hog Meat and Hoe Cake: Food Supply in the Old South, 1840–1860* (Carbondale, Ill., 1972).

46. It is well known that in the antebellum period country storekeepers did play a role in the marketing of cotton and the advancement of credit, mainly for small producers. See Lewis E. Atherton, *The Southern Country Store, 1800–1860* (Baton Rouge, La., 1949); and Harold D. Woodman, *King Cotton and His Retainers: Financing and Marketing of the Cotton Crop of the South, 1800–1925* (Lexington, Ky., 1968), esp. 76–83. On the lock-in mechanism and the postbellum shift to increasing reliance on cotton, see Ransom and Sutch, *One Kind of Freedom*, 149–170. Critiques of the lock-in model can be found in Claudia Goldin, "'N' Kinds of Freedom: An Introduction to the Issues," *Explorations in Economic History* 16 (Jan. 1979): 8–30; and in Peter Temin, "Freedom and Coercion: Notes on the Analysis of Debt Peonage in *One Kind of Freedom*," ibid., 56–63. An important start in analyzing surviving antebellum rental contracts has been made by Reid, "Antebellum Southern Rental Contracts."

47. On the cotton-intensity ratio, see also chapter 3.

48. The differences in the weight of cotton bales are discussed in chapter 7.

CHAPTER 4: THE DELINEATION OF AGRICULTURAL REGIONS IN GEORGIA

1. U.S. Census Office, Tenth Census [1880], *Report on Cotton Production in the United States*, Part II, by Eugene W. Hilgard et al., *Report on the Cotton Production of the State of Georgia* by R. H. Loughridge (Washington, D.C., 1884) [hereinafter cited as *1880 Cotton Report (Georgia)*].

2. Dade, the only county represented in region I, will play no part in our analysis.

3. For classic statements of these regional differences, see Frank L. Owsley, *Plain Folk of the Old South* (Chicago, 1965 [first published, 1949]); Morton Rothstein, "The Antebellum South as a Dual Economy: a Tentative Hypothesis," *Agricultural History* 41 (Oct. 1967): 373–83; and Eugene D. Genovese, "Yeoman Farmers in a Slaveholders' Democracy," *Agricultural History* 49 (April 1975): 331–42.

4. After examining rank-order distributions on several leading variables, Ransom and Sutch reassigned Brooks from the Oak, Hickory and Pine Uplands to the Limesink. They also reassigned several counties of the Limesink—stretching from Mitchell and Baker to Early—to the Central Cotton Belt. Ransom and Sutch, *Economic Regions of the South*, Working Paper No. 3, Southern Economic History Project, appendix A.

5. Donald Steila, *The Geography of Soils: Formation, Distribution, and Management* (Englewood Cliffs, N.J., 1976), 143.

6. Much of the historical literature assumes too easily that cotton was *necessarily* highly depleting of southern soils. There is no doubt that cotton did deplete the soils on

which they grew, but such depletion was a function of poor soil management, which was a pervasive characteristic of southern antebellum agriculture. The *1880 Cotton Report (Georgia)*, 60, makes it quite clear that cotton takes less from soil than either wheat or corn, if (and only if) seed husks are returned to the soil as fertilizer. According to the *Report*, corn stalks were almost always returned, whereas antebellum cotton production in general never returned seeds and virtually threw them away.

7. Terrell in the southwest, another Ransom and Sutch county, is also class 4 in 1860, that is, it has tenants but they are reported with full values on schedule IV. It is now apparent that such counties would have been valuable additions to the sample. But when we were actually selecting our counties, at the outset of our research and well before we had developed the short method, they seemed poor candidates for inclusion in a study of tenancy.

CHAPTER 5: THE FORMS OF TENANCY IN ANTEBELLUM GEORGIA

1. Harold D. Woodman, "Post-Civil War Southern Agriculture and the Law," *Agricultural History* 53 (Jan. 1979): 325–26; and Steven Hahn, *The Roots of Southern Populism: Yeomen Farmers and the Transformation of the Georgia Upcountry, 1850–1890* (New York and Oxford, 1983), 162–63.

2. Charles S. Mangum, Jr., *The Legal Status of the Tenant Farmer in the Southeast* (Chapel Hill, N.C., 1952), 12–13, 22–23. See also Woodman, "Post-Civil War Southern Agriculture and the Law," 310–37; and Robert Preston Brooks, *The Agrarian Revolution in Georgia, 1865–1912* (Westport, Conn., 1970 [first published, 1914]), 52–55, 65–67, 79.

3. See, for example, Robert Higgs, "Patterns of Farm Rental in the Georgia Cotton Belt, 1880–1900," *The Journal of Economic History* 34 (June 1974): 468–82; and Joseph D. Reid, Jr., "White Land, Black Labor, and Agricultural Stagnation: The Causes and Effects of Sharecropping in the Postbellum South," *Explorations in Economic History* 16 (Jan. 1979): 31–55.

4. For further discussion of tenancy rates in 1880, see below, chapter 7.

5. Brooks, *Agrarian Revolution in Georgia*, 55, 79.

6. Roger L. Ransom and Richard Sutch, *One Kind of Freedom: The Economic Consequences of Emancipation* (Cambridge, 1977), 88. For a good discussion of the wide variety of labor arrangements that prevailed in South Carolina between 1865 and 1868, see Ralph Shlomowitz, "The Origins of Southern Sharecropping," *Agricultural History* 53 (July 1979): 557–75.

7. *Southern Cultivator* 26 (Feb. 1868): 61–62; 25 (Dec. 1867): 364–65; 26 (Jan. 1868), 12–13; and May 1868, 133.

8. No one has as yet undertaken a full study of the conditions that led to the attempts at more precise legal definitions of tenancy arrangements in the 1870s.

9. Joseph D. Reid, Jr., "Antebellum Southern Rental Contracts," *Explorations in Economic History* 13 (Jan. 1976): 69–83. Hayward is an upland county and grew no cotton. The fifty-nine leases were made by two landlords between 1821 and the Civil War.

10. Hahn, *Roots of Southern Populism*, 64–69. Lewis Cecil Gray has also commented on the use of tenancy for land clearance. *History of Agriculture in the Southern United States* (Gloucester, Mass., 1958 [first published, 1932]), vol. 2, 646–47. According to Enoch Marvin Banks, sharecropping was "a post-bellum product," but renting on thirds and

fourths began before the war. *The Economics of Land Tenure in Georgia* (New York, 1905), 82.

11. Marjorie Mendenhall Applewhite, "Sharecropper and Tenant in the Courts of North Carolina," *North Carolina Historical Review* 31 (April 1954), 136–38.

12. Quoted in Brooks, *Agrarian Revolution in Georgia*, 67. Emphasis added by Brooks.

13. Banks, *Economics of Land Tenure in Georgia*, 81–82.

14. Antebellum statute law in Georgia recognized the normal landlord-tenant (*i.e.*, renter) relationship in which "the owner of lands grants to another simply the right to possess and enjoy the use of such land, either for a fixed time or at the will of the grantor ..." and also provided for the payment of rent in kind. The statutes on hiring provided that payment for labor could take the form of shares of the goods produced, thus allowing in principle for the legal existence of sharecropping. R. H. Clark, T. R. R. Cobb, and D. Irwin, *The Code of the State of Georgia* (Atlanta, 1861), 408, 436–38.

15. We have examined the complete files of the *Southern Cultivator* for the antebellum years beginning with its first issue in 1843. See also Gray, *History of Agriculture*, vol. 2, 646.

16. See above, chapter 2.

17. Possibly some class 3 counties, which were not analyzable by the short method, excluded tenants from schedule IV.

18. Since we examined only selected pages of schedule I for most counties examined by the short method, we may have overlooked tenancy candidates in Monroe and Telfair.

19. Seddie Cogswell, Jr., *Tenure, Nativity and Age as Factors in Iowa Agriculture, 1850–1880* (Ames, Iowa, 1975), 7–8. See also Allan G. Bogue, *From Prairie to Cornbelt: Farming on the Illinois and Iowa Prairies in the Nineteenth Century* (Chicago, 1963), 63–64; and Merle Curti, *The Making of an American Community: A Case Study of Democracy in a Frontier Community* (Stanford, Calif., 1959), 59–60.

20. We have shown that not all types b and c were first-year farmers. In Houston County, for example, most, but not all, types b and c had nonagricultural occupations and seem to have been town dwellers whose livestock holdings the enumerator recorded on schedule IV. But in Spalding County, examined by the short method, all but one type b and one type c were designated by the enumerator as a first-year farmer. One farmer with arable production was so designated. Some first-year farmers with no arable will appear in our data as type a if their livestock holdings exceeded the maximum limit for type b. But if all first-year farmers were entered on schedule IV in any county, almost certainly some of them would appear as type b or c. In Spalding these types accounted for almost a third of the first-year farmers. See chapter 2.

21. These counties are Baker, Baldwin, Clay, Emanuel, Greene, Gwinnett, Thomas, Wilkes, and Worth. What could have happened to the first-year farmers in Echols is a mystery. It seems unlikely that there were none in the county.

22. The few type 4 farmers not appearing in the four counties could have been retired farmers, farmers whose holdings appeared under the name of a manager on schedule IV, or farmers resident in the county of enumeration but owning their farm property elsewhere.

23. Easterlin et al., "Farms and Farm Families in Old and New Areas: The Northern States in 1860," in *Family and Population in Nineteenth-Century America*, Tamara K.

Hareven and Maris A. Vinovskis, eds. (Princeton, N.J., 1978), 54–56. Harold Woodman claims that sharecropping was confined to the South but cites no evidence for this claim. "Post-Civil War Southern Agriculture," 325–26. Paul W. Gates has found that in 1860 the census enumerator for Bureau County, a major wheat, corn, and livestock county in Illinois, explicitly listed 2,433 "farmers," 288 "tenants," 81 "agents," and 14 persons as "renting land." "Tenants," according to Gates, "were defined [by the enumerator] as persons who owned all the stock and paid either a cash or a share rent; agents were defined as persons who owned no stock but were provided with stock by the landlord and in turn paid a share rent." Although the so-called agents "paid a share rent," the furnishing of supplies suggests the classic sharecropping pattern. In any case, tenancy arrangements in this midwestern county were complex. Gates also found similar definitions in Kankakee County, Illinois, where, however, "incomplete data" yielded 136 tenants and only one agent. *The Farmer's Age: Agriculture, 1815–1860* (New York, 1968 [first published, 1960]), 198.

24. The high farm laborer ratio in Wilkes is explained by the fact that the enumerator designated overseers as farm laborers. See above, chapter 5.

25. *1880 Cotton Report (Georgia)*, 172.

26. Quoted in Gray, *History of Agriculture*, vol. 1, 468.

27. Houston reported six type 2 tenants with farm value only, and Twiggs reported one with farm value only. Possibly these cases were errors. Habersham and Harris also reported a small number of type 3 possible tenants, *i.e.*, entries with no real property on schedule I and with at least improved acres and farm value reported on schedule IV. We have already suggested that type 3 entries may have been cases of exclusively tenanted holdings when the proprietor did not operate a farm on his own behalf in the county. They could also be errors resulting from the enumerator's failure properly to assign a value for real property on schedule I.

28. The census definition of personal property was meant to be all-inclusive except for real estate. The instructions specifically mentioned livestock as an example of personal property.

29. See above, chapter 3.

30. Mangum, *Legal Status of the Tenant Farmer*, 12–13, 23. Emphasis has been added in the quotation.

31. The generalizations in this paragraph are based on data in appendix F.

32. Gilmer also reported two tenants in standard convention and three with improved acres only. It is not possible to determine with any certainty whether missing unimproved acres constituted a separate tenancy convention or simply meant that the tenant possessed no unimproved acres.

33. Those counties having two or more enumerators who adopted different procedures are not identified in appendix A. In a few counties different enumerators did use different conventions, which do not reflect real differences in tenancy forms. In Jackson County, for example, one of two enumerators consistently employed the standard convention, while the other consistently recorded improved acres only and wrote "tenant" in the unimproved column.

34. The status of the large number of "croppers" who were not heads of household in Walton County is uncertain. Most of these croppers were apparent sons with the same surname as the household heads (who were almost always "farmers"), but a sub-

stantial number were apparent boarders with different surnames. That these croppers were not merely casual laborers is attested by the fact that some of these same households also reported individuals designated "farm laborer" and "day laborer." It may be that cropper sons farmed a portion of the family holdings with supplies furnished by their fathers. If so, their position was analogous to other croppers, who were employees of the proprietor. This pattern suggests, moreover, that in other counties some number of farmers' sons who were conventionally designated "farmer," "farm laborer," or given no occupation may also have separately farmed parts of the family holdings in a cropper-like arrangement.

35. In all three counties a few individuals designated "tenant," "renter," or "cropper" reported real property. It may be that they were tenants on parts of their holdings. Some of these cases could also have been enumerator errors.

36. See appendix A. The sixteen entries in Madison reporting improved acres and farm value with "let" noted in the unimproved acreage column may not have been tenants, since they all reported real property on schedule I that was equal to their farm value, and some of them reported substantial production. Possibly the names recorded for these entries were proprietors who had let their holdings to tenants who, perhaps because they were sharecroppers (despite a misuse of the term *let*), were not separately entered on schedule IV. The excess of tenancy candidates on schedule I over the number of standard convention type a entries in Madison may be accounted for by types b and c, though the greater number of these types could have been town dwellers with small numbers of livestock.

37. We found no schedule I "tenants" on schedule IV, but they might be found among the majority of schedule IV entries that, by the short method, we did not link to schedule I.

38. In examining ninety schedule IV entries in Paulding by the short method we found four type 3 records.

39. We wish to caution the reader again that generalizations about schedule IV entries in these three counties are based only on short-method estimates.

40. Hahn, *Roots of Southern Populism*, 64–69.

41. Type 2 entries in Worth County were not entered in a schedule IV convention, but the enumerator identified them as tenants on that schedule.

42. The 1870 census, which did not separately classify tenants, defined farms in the same way as the 1880 census.

CHAPTER 6: THE SPATIAL DISTRIBUTION OF TENANCY IN 1860

1. One could lower level I estimates by introducing type 4 records into the denominator as possible landlords within the county; however, to introduce type 4 records without introducing 5 seems unwise.

2. Patterns of reporting tenancy in Fannin and Gilmer are quite different from those in Towns and should caution against easy generalizations regarding modes of tenancy in the mountains and elsewhere, however.

3. For a definition of short-method "classes," see chapter 2.

4. For a more complete report of short method findings, see appendix B.

5. Owsley does not fully explain his methods, but there are passages that tend to

confirm our conclusions. Frank Owsley, *Plain Folk of the Old South* (Chicago, 1965 [first published, 1949]), 158, 160, 191, 215.

6. We counted 123 nonfarming occupations among Harris farm operators on schedule IV. Therefore the discrepancy cannot be explained by Owsley's having excluded nonfarming occupations. There is no evidence here or elsewhere that he did so.

7. We have not worked directly with county tax lists. According to Owsley (*Plain Folk*, 150 n. 1, 184), "the taxbooks failed to indicate what a person's vocation was, or whether he was the head of a family, and thus made them less valuable as a source in studying land tenure of the agricultural population." But on the same page (150) he also expresses the opinion that tax lists are more accurate, since the tax officer was collecting money and not merely names. A large proportion of households encountered on schedule I in most counties are not found on schedule IV and report nonfarming occupations. If it is true that tax lists fail to designate occupation, then they can be of little value for estimating farm tenancy. At best they would estimate the percentage of landed proprietors among all persons assessed for taxes of any kind; and if "head of household" is not included as a designation, then further contamination is introduced. Tenancy rates produced by such records would yield percentages well above our own.

8. Donghyu Yang has constructed 1860 tenancy estimates from the Bateman-Foust sample of northern farms for sixteen northern states. Both the spatial patterns and levels of northern tenancy rates bear some resemblance to those we have found in Georgia and add to their plausibility. The northern rates, when compared with our own, also open new questions of North-South differences that need to be explored. His estimates are not fully comparable with our own, since he has excluded from his counts all farms on schedule IV whose head of household failed to report his occupation as "farmer" on schedule I; additionally, Yang has not standardized for types b and c. In every other respect his column B of table 2 is equivalent to our level II, while his column B plus column C is equivalent to our level III. He finds a 19.4 percent tenancy rate at level III for the northern sample as a whole. Individual states vary from 7 percent in Connecticut to 48.5 percent in Maryland. In general the older lands of New England remain below 10 percent, with rates tending to rise as one moves onto the newer lands of the west. Kansas and Minnesota each achieve rates in the thirties, for example. It is clear that there are local peculiarities in the modes of reporting, however. Michigan achieves an extraordinarily low level III rate of 5.1 percent; but 34.3 percent of the analyzed records are type 4, which is wholly out of line with all other northern states and with our findings in Georgia. Donghyu Yang, "Notes on the Wealth Distribution of Farm Households in the United States, 1860: A New Look at Two Manuscript Census Samples," *Explorations in Economic History* 21 (Jan. 1984): 88–102. We are grateful to Dr. Yang for permitting us to see his paper before publication.

9. The district lies in approximately two halves, the northern half entirely in the Sand Hills and the southern half on superior lands that in adjoining districts were heavily under cotton production by large holders. District 356 contained two type 1a proprietors who accounted for 42 percent of the total acres of the district, 47.4 percent of the improved acreage, and 36.8 percent of the cotton. In other words these two type 1a proprietors were capable of accounting for virtually the entire southern portion of the district, and their production figures suggest they were not heavily reliant on tenants. The high tenancy rate for the district must therefore be essentially correct for the

Notes   265

Sand Hills portion, since the inclusion of two in the denominator would not significantly affect the calculation.

10. When y = tenancy rate per district at level IV and x = farm value per total acre in Baldwin county (N = 7), $r = -.891$ and is significant at .99 when all districts are included ($r = -.711$ and is significant at .80 when district 320, containing Milledgeville, is deleted); when x = mean improved acres, $r = -.205$ and is not significant at .80 ($r = -.785$ and is significant at .90 when district 320 is deleted); when x = mean type 1a personal, $r = -.289$ and is not significant at .80 ($r = -.833$ and is significant at .90 when districts 320 and 321 are deleted and district 318 is revised to exclude the record of one planter reporting personal property in excess of one million dollars); when x = improved acres per bale for all landholders, $r = -.062$ and is not significant at .80 ($r = .642$ and is significant at .80 when district 320 is deleted); when x = improved acres per bale for cotton producers only, $r = -.189$ and is not significant at .80 ($r = .240$ and is not significant at .80 when district 320 is deleted); when y = improved acres per bale for all landholders and x = mean improved acres, $r = -.759$ and is significant at .95 ($r = -.727$ and is significant at .80 when district 320 is deleted). When y = tenancy rate per district at level III and x = farm value per total acre in Harris County (N = 13), $r = -.223$ and is not significant at .80 ($r = -.252$ and is not significant at .80 when banner district 934 is deleted); when x = mean improved acres, $r = -.374$ and is not significant at .80 ($r = -.627$ and is significant at .95 when district 934 is deleted); when x = mean type 1a personal property, $r = -.244$ and is not significant at .80 ($r = -.601$ and is significant at .95 when district 934 is deleted); when x = improved acres per bale for all landholders, $r = -.430$ and is significant at .80 ($r = -.438$ and is significant at .80 when district 934 is deleted); when x = improved acres per bale of cotton producers only, $r = -.505$ and is significant at .90 ($r = -.525$ and is significant at .90 when district 934 is deleted); when y = improved acres per bale for all landholders and x = mean improved acres, $r = .315$ and is not significant at .80 ($r = .417$ and is significant at .80 when district 934 is deleted). When y = tenancy rate per district at level III and x = farm value per total acre in Twiggs County (N = 10), $r = -.129$ and is not significant at .80 ($r = -.022$ and is not significant at .80 when banner district 396 is deleted); when x = mean improved acres, $r = .046$ and is not significant at .80 ($r = -.388$ and is not significant at .80 when district 396 is deleted); when x = mean type 1a personal, $r = .007$ and is not significant at .80 ($r = -.434$ and is not significant at .80 when district 396 is deleted); when x = improved acres per bale for all landowners, $r = -.725$ and is significant at .95 ($r = -.689$ is significant at .95 when district 396 is deleted); when x = improved acres per bale for cotton producers only, $r = -.664$ and is significant at .95 ($r = -.617$ and is significant at .90 when district 396 is deleted); when y = improved acres per bale for all landholders and x = mean improved acres, $r = -.340$ and is not significant at .80 ($r = -.082$ and is not significant at .80 when district 396 is deleted). Multiple regressions were not attempted when the data were found to be defective in the way they were assigned to districts. Great caution should be employed in interpreting these statistics.

11. For a more extended discussion of county farm acreage as reported in the 1860 census, see chapter 2, above.

12. We have been careful to select only militia districts not immediately adjacent either to counties created after 1860 or to other militia districts also created after that date. Our impression is that while militia-district boundaries undoubtedly changed

over the years, the changes were only significant when they were coupled with the creation of new counties and districts.

13. We have generally resisted more sophisticated statistical analysis because of the doubts we currently have about the reliability of the data. Until more work is completed on the census as a data source, we advise restraint. We also caution against any easy use of statistics offered in footnote 10 above; they are not presented in tabular form precisely to discourage such use.

CHAPTER 7: THE TRANSITION TO 1880

1. Roger L. Ransom and Richard C. Sutch, *Economic Regions of the South in 1880*, Working Paper No. 3, Southern Economic History Project, 30–31, 35–38.

2. Writing of the hill country of Alabama, Jonathan M. Wiener argues that before the Civil War "there had been no planter class in the hills, a region of small farmers. The rise of a new class of merchant-landlords entailed the fall of the white yeoman farmers to tenant status" in the decade or so following the war. "The class structure of the hills came increasingly to resemble that of the black belt . . . [and] the white farmers in the hills were forced into tenancy by the merchant elite's monopoly on credit and increasing ownership of land." *Social Origins of the New South: Alabama, 1860–1885* (Baton Rouge, La., 1978), 93.

3. "Family tenancy apparently emerged in the Cotton South as an almost unprecedented form of labor organization. It appears that, before the Civil War, almost all cotton farms were operated by owners or their managers. According to the studies made by Frank Owsley and his students of the manuscript censuses of 1850 and 1860, approximately 80 percent of all farm operators owned real estate. This figure has been interpreted as a lower bound on the percentage of farms that were owner-operated. In all probability the true figures would be closer to 90 percent. In any event, whatever the extent of prewar land renting, we can be confident that most if not all of it involved the leasing of plantations or medium-scale farms that were operated by the leaseholder on the plantation system. There does not seem to have been any leasing of small land parcels to family tenants or sharecroppers aside from a very few isolated cases. Lewis Gray has argued that the scarcity of such references is strong evidence of the absence of family tenancy as a form of labor organization before the Civil War. Tenancy did not reflect a prewar form of agricultural organization; rather it seems to have emerged as the most practical of the many experimental systems tried by determined groups of planters who sought alternatives to the plantation system as early as 1865." Ransom and Sutch, *One Kind of Freedom: The Economic Consequences of Emancipation* (Cambridge, 1977), 88.

4. Ransom and Sutch (*One Kind of Freedom*, 105) have made a similar point, but from a perspective that assumed no significant white tenancy before the Civil War and therefore failed to appreciate the importance to white yeomen and tenant families of losing a former monopoly. If poor to middling whites "fell" into anything, it was into a morass of direct competitive relations with blacks to which they were unaccustomed.

5. We have no firm estimates of tenancy for 1860 in the coastal counties, but brief attempts to apply the short method left us with the clear impression that they were very low, perhaps 10 percent or less in some counties. The lower coastal counties therefore

probably doubled their rates by 1880, but the rise would seem to have been steeper as one moves up the coast.

6. See, for example, U.S. Census Office, Tenth Census [1880], *Report on Cotton Production in the United States*, Part II, by Eugene W. Hilgard et al., *Report on the Cotton Production of the State of Georgia* by R. H. Loughridge (Washington, D.C., 1884) [hereinafter cited as *1880 Cotton Report (Georgia)*].

7. See notes to appendix G.

8. Robert Preston Brooks, *The Agrarian Revolution in Georgia, 1865–1912* (Westport, Conn., 1970 [first published, 1914]), 79. For a fuller discussion of bases for enumerator confusion, see chapter 5, above.

9. It should also be noted, of course, that Georgia undoubtedly departed from the mean experienced by other southern states. Such a surrogate rate is automatically adjusted downward for Civil War losses.

10. Direct evidence that tenants in 1881 were being employed for clearance of previously wooded portions of a holding is provided by a map of the Barrow plantation, located in Oglethorpe County on the older lands of the eastern Piedmont (Ransom and Sutch, *One Kind of Freedom*, 72). It would indeed be surprising if at least some tenants in all regions were not engaged in some amount of clearance, however. The problem lies at a more general level.

11. *1880 Cotton Report (Georgia)*, 56, 59.

12. Some twenty Georgia counties in 1880 used the Scofield press, which produced bales weighing 600 pounds each (ibid., 170).

13. James C. Bonner, *A History of Georgia Agriculture, 1732–1860* (Athens, Ga., 1964), 192. Lewis Cecil Gray, *History of Agriculture in the Southern United States to 1860* (Gloucester, Mass., 1958 [first published, 1933]), vol. 2, 705, recounts the slow rise in mean bale weight that occurred throughout the first half of the century, noting regional differences.

14. Steven Hahn incorrectly characterizes the upper Piedmont counties after the Civil War as having "spiraling tenancy" in contrast to the counties of the old Plantation Belt. *The Roots of Southern Populism: Yeomen Farmers and the Transformation of the Georgia Upcountry, 1850–1890* (New York and Oxford, 1983), 4.

Bibliography

Applewhite, Marjorie Mendenhall. "Sharecropper and Tenant in the Courts of North Carolina." *North Carolina Historical Review* 31 (April 1954): 134–49.

Atherton, Lewis E. *The Southern Country Store, 1800–1860*. Baton Rouge, La., 1949.

Banks, Enoch Marvin. *The Economics of Land Tenure in Georgia*. New York, 1905.

Bode, Frederick A., and Donald E. Ginter. "A Critique of Landholding Variables in the 1860 Census and the Parker-Gallman Sample." *Journal of Interdisciplinary History* 15 (Autumn 1984): 277–95.

Bogue, Allan G. *From Prairie to Cornbelt: Farming on the Illinois and Iowa Prairies in the Nineteenth Century*. Chicago, 1963.

Bonner, James C. *A History of Georgia Agriculture, 1732–1860*. Athens, Ga., 1964.

———. "Profile of a Late Ante-Bellum Community." *American Historical Review* 49 (July 1944): 663–80.

Brooks, Robert Preston. *The Agrarian Revolution in Georgia, 1865–1912*. Westport, Conn., 1970 (first published, 1914).

Campbell, Randolph B. "Planters and Plainfolk: Harrison County, Texas, as a Test Case, 1850–1860." *Journal of Southern History* 40 (August 1974): 369–98.

———, and Richard G. Lowe. *Wealth and Power in Antebellum Texas*. College Station, Texas, 1977.

Clark, Blanche Henry. *The Tennessee Yeomen, 1840–1860*. New York, 1971 (first published, 1942).

Clark, R. H., T. R. R. Cobb, and D. Irwin. *The Code of the State of Georgia*. Atlanta, 1861.

Cogswell, Seddie, Jr. *Tenure, Nativity and Age as Factors in Iowa Agriculture, 1850–1880.* Ames, Iowa, 1975.

Coles, Harry L., Jr. "Some Notes on Slave Ownership and Land Ownership in Louisiana, 1850–1860." *Journal of Southern History* 9 (August, 1943): 381–94.

Curti, Merle. *The Making of an American Community: A Case Study of Democracy in a Frontier Community.* Stanford, Calif., 1959.

David, Paul A., et al. *Reckoning with Slavery: A Critical Study in the Quantitative History of American Negro Slavery.* New York, 1976.

Davis, Charles S. *The Cotton Kingdom in Alabama.* Philadelphia, 1974 (first published, 1939).

DeCanio, Stephen J. *Agriculture in the Postbellum South: The Economics of Production and Supply.* Cambridge, Mass., 1974.

Easterlin, Richard A., et al. "Farms and Farm Families in Old and New Areas: The Northern States in 1860." In *Family and Population in Nineteenth-Century America*, edited by Tamara K. Hareven and Maris A. Vinovskis. Princeton, N.J., 1978, 22–84.

Fogel, Robert William, and Stanley L. Engerman. *Time on the Cross.* Vol. 1. *The Economics of American Negro Slavery.* Vol. 2. *Evidence and Methods—A Supplement.* Boston and Toronto, 1974.

Foust, James D. "The Yeoman Farmer and the Westward Expansion of U.S. Cotton Production." Ph.D. dissertation, University of North Carolina, 1967. Reprinted by Arno Press, 1976.

———, and Dale E. Swan. "Productivity and Profitability of Antebellum Slave Labor: A Micro-Approach." *Agricultural History* 44 (Jan. 1970): 39–62.

Gallman, Robert E. "Self-Sufficiency in the Cotton Economy of the Antebellum South." *Agricultural History* 44 (Jan. 1970): 5–23.

Gates, Paul W. *The Farmer's Age: Agriculture, 1815–1860.* New York, 1968 (first published, 1960).

Genovese, Eugene D. *The Political Economy of Slavery: Studies in the Economy and Society of the Slave South.* New York, 1965.

———. "Yeoman Farmers in a Slaveholders' Democracy." *Agricultural History* 49 (April 1975): 331–42.

Georgia. Office of the Comptroller General. *Annual Report of the Comptroller General, of the State of Georgia, Made to the Governor, October, 1857.* Columbus, n.d.

———. *Annual Report of the Comptroller General . . . October 20, 1858.* Columbus, 1858.

———. *Annual Report of the Comptroller General . . . October 20, 1859.* Milledgeville, 1859.

———. *Annual Report of the Comptroller General . . . October 20, 1860.* Milledgeville, 1860.

Goldin, Claudia, "'N' Kinds of Freedom: An Introduction to the Issues." *Explorations in Economic History* 16 (Jan. 1979): 8–30.
Gray, Lewis Cecil. *History of Agriculture in the Southern United States to 1860*, 2 vols. Gloucester, Mass., 1958 (first published, 1933).
Hahn, Steven. *The Roots of Southern Populism: Yeomen Farmers and the Transformation of the Georgia Upcountry, 1850–1890*. New York and Oxford, 1983.
Heath, Milton Sydney. *Constructive Liberalism: The Role of the State in Economic Development in Georgia to 1860*. Cambridge, Mass., 1954.
Higgs, Robert. *Competition and Coercion: Blacks in the American Economy, 1865–1914*. Cambridge, 1977.
———. "Patterns of Farm Rental in the Georgia Cotton Belt, 1880–1900." *Journal of Economic History* 34 (June 1974): 468–82.
Hilliard, Sam Bowers. *Hog Meat and Hoe Cake: Food Supply in the Old South, 1840–1860*. Carbondale, Ill., 1972.
Huffman, Frank Jackson, Jr. "Old South, New South: Continuity and Change in a Georgia County, 1850–1880." Unpublished Ph.D. dissertation, Yale University, 1974.
Johnson, Guion Griffis. *Ante-Bellum North Carolina: A Social History.* Chapel Hill, N.C., 1937.
Linden, Fabian. "Economic Democracy in the Slave South: An Appraisal of Some Recent Views." *Journal of Negro History* 31 (April 1946): 140–89.
Mangum, Charles S., Jr. *The Legal Status of the Tenant Farmer in the Southeast.* Chapel Hill, N.C., 1952.
Mendenhall, Marjorie Stratford. "The Rise of Southern Tenancy." *Yale Review*, n.s. 27 (Sept. 1937): 110–29.
Owsley, Frank L. *Plain Folk of the Old South*. Chicago, 1965 (first published, 1949).
——— and Harriet C. Owsley. "The Economic Basis of Society in the Late Ante-Bellum South." *Journal of Southern History* 6 (Feb. 1940): 24–25.
———. "The Economic Structure of Rural Tennessee, 1850–1860." *Journal of Southern History* 8 (May 1942): 161–82.
Phillips, Ulrich Bonnell. *American Negro Slavery: A Survey of the Supply, Employment and Control of Negro Labor as Determined by the Plantation Regime*. Baton Rouge, La., 1966 (first published, 1918).
Range, Willard. *A Century of Georgia Agriculture, 1850–1950*. Athens, Ga., 1954.
Ransom, Roger L., and Richard Sutch. *Economic Regions of the South*. Working Paper No. 3. Southern Economic History Project.
———. *One Kind of Freedom: The Economic Consequences of Emancipation*. Cambridge, 1977.
Reid, Joseph D., Jr. "Antebellum Southern Rental Contracts." *Explorations in Economic History* 13 (Jan. 1976): 69–83.

272 Bibliography

———. "White Land, Black Labor, and Agricultural Stagnation: The Causes and Effects of Sharecropping in the Postbellum South." *Explorations in Economic History* 16 (Jan. 1979): 31–55.

Rothstein, Morton. "The Antebellum South as a Dual Economy: A Tentative Hypothesis." *Agricultural History* 41 (Oct. 1967): 373–83.

Shlomowitz, Ralph. "The Origins of Southern Sharecropping." *Agricultural History* 53 (July 1979): 557–75.

Soltow, Lee. *Patterns of Wealthholding in Wisconsin since 1850.* Madison, Wis., 1971.

Southern Cultivator. Vols. 1–27 (1843–69).

Stampp, Kenneth M. *The Peculiar Institution: Slavery in the Antebellum South.* New York, 1956.

Swierenga, Robert P. "Quantitative Methods in Rural Landholdings." *Journal of Interdisciplinary History* 13 (Spring 1983): 787–808.

Temin, Peter. "Freedom and Coercion: Notes on the Analysis of Debt Peonage in *One Kind of Freedom.*" *Explorations in Economic History* 16 (Jan. 1979): 56–63.

U.S. Census Office, Eighth Census [1860]. *Agriculture of the United States in 1860, Compiled from the Original Returns of the Eighth Census.* Washington, D.C., 1864.

———. *Instructions to U.S. Marshalls. Instructions to Assistants.* Washington, D.C., 1860.

———. *Population of the United States in 1860, Compiled from the Original Returns of the Eighth Census.* Washington, D.C., 1864.

———. Schedule 1, Free Inhabitants (Georgia). Microfilm of manuscript enumerator returns in the U.S. Archives.

———. Schedule IV, Productions of Agriculture (Georgia). Microfilm of manuscript enumerator returns in Duke University Library.

U.S. Census Office. Tenth Census [1880]. *Report on Cotton Production in the United States Also Embracing Agricultural and Physico-Geographical Descriptions of the Several Cotton States and California* by Eugene W. Hilgard, et al. 2 vols. Washington, D.C., 1884.

———. *Report on the Production of Agriculture in the United States at the Tenth Census (June 1, 1880).* Washington, D.C., 1883.

———. *Statistics of the Population of the United States at the Tenth Census (June 1, 1880).* Washington, D.C., 1883.

Weaver, Herbert. *Mississippi Farmers, 1850–1860.* Gloucester, Mass., 1968 (first published, 1945).

White, George. *Statistics of the State of Georgia: Including an Account of the Natural, Civil, and Ecclesiastical History; Together with a Particular Description of Each County . . . and a Correct Map of the State.* Spartanburg, S.C., 1972 (first published, 1849).

Wiener, Jonathan M. "Class Structure and Economic Development in the American South, 1865–1955." *American Historical Review* 84 (Oct. 1979): 970–92.

———. *Social Origins of the New South: Alabama, 1860–1885.* Baton Rouge, La., 1978.

Winters, Donald L. *Farmers without Farms: Agricultural Tenancy in Nineteenth Century Iowa.* Westport, Conn., 1978.

Woodman, Harold D. *King Cotton and His Retainers: Financing and Marketing of the Cotton Crop of the South, 1800–1925.* Lexington, Ky., 1968.

———. "Post-Civil War Southern Agriculture and the Law." *Agricultural History* 53 (Jan. 1979): 310–37.

———. "Sequel to Slavery: The New History Views the Postbellum South." *Journal of Southern History* 43 (Nov. 1977): 523–54.

Woodward, C. Vann. *Origins of the New South, 1877–1913.* Baton Rouge, La., 1951.

Wright, Carroll D. *The History and Growth of the United States Census, Prepared for the Senate Committee on the Census.* Washington, D.C., 1900.

Wright, Gavin. "'Economic Democracy' and the Concentration of Agricultural Wealth in the Cotton South, 1850–1860." *Agricultural History* 44 (Jan. 1970): 63–85.

———. *The Political Economy of the Cotton South: Households, Markets, and Wealth in the Nineteenth Century.* New York, 1978.

———. "The Strange Career of the New Southern Economic History." *Reviews in American History* 10 (Dec. 1982): 164–80.

Yang, Donghyu. "Notes on the Wealth Distribution of Farm Households in the United States, 1860: A New Look at Two Manuscript Census Samples." *Explorations in Economic History* 21 (Jan. 1984): 88–102.

Index

Agricultural ladder, 6, 46, 254 (n. 3)
Alabama, 136–37, 203, 266 (n. 2)
Appling County, Georgia, 163
Arkansas, 105

Baker County, Georgia, 4, 16, 22, 31–32, 35, 38, 49, 51–52, 56–57, 61, 68–72, 81, 89, 98, 100, 115–37, 144–45, 150–51, 173, 255 (n. 17), 258 (n. 38), 259 (n. 4), 261 (n. 21)
Baldwin County, Georgia, 4, 16, 22–24, 31–32, 35, 38, 49, 51–52, 56, 60–62, 68–71, 89, 98, 100–101, 108, 115–45, 151–52, 160, 258 (n. 38), 261 (n. 21)
Banks, Enoch Marvin, 7, 260 (n. 10)
Bibb County, Georgia, 51, 61, 160
Bogue, Allan G., 39, 253 (n. 40)
Bonner, James C., 169–70, 253 (n. 37)
Brooks, Robert Preston, 92
Brooks County, Georgia, 82, 156, 259 (n. 4)
Bryan County, Georgia, 84
Bulloch County, Georgia, 163
Butts County, Georgia, 250 (n. 13)

Camden County, Georgia, 83–84
Campbell, Randolph, 40–41, 253 (n. 46), 253–54 (n. 47)
Carroll County, Georgia, 42–43, 93, 254 (n. 52)
Census of 1880, 11

Census, manuscript (1860), 249 (n. 2); schedule IV (agriculture), 6, 8–9, 11–46, 50–67, 72–73, 96–113, 141–42; schedule I (population), 8, 11–57, 67–72, 96–113, 141–43; agent, manager, or overseer, 13, 19, 21–24, 126, 250 (n. 15), 250–51 (n. 16), 262 (nn. 23, 24); cotton bales, 13, 26, 45, 72–73, 142–43, 167–73, 178, 184; crops and livestock, 13, 15–20, 142–43; farm value, 13, 16, 21–33, 45, 50, 57–67, 142–43; improved acres, 13, 21–33, 45, 57–67, 142–43, 204–6; occupation, 13–15, 18–19, 45–50, 249 (n. 6), 251 (n. 21), 254 (n. 49); personal property, 13, 22, 25–26, 45, 67–72, 104–5, 142–43, 207–22; real property, 13, 15, 21–26, 45, 50–57, 108; unimproved acres, 13, 21–33, 45, 57–67, 142–43, 204–6; defective variables, 25–26, 256 (n. 19). *See also* Laborers, farm and other
Charlton County, Georgia, 83
Chatham County, Georgia, 84, 160
Chattooga County, Georgia, 131
Clark, Blanche Henry, 33–35, 37, 136–37, 181, 203
Clarke County, Georgia, 41–42
Classes 1 to 5 (short method), 28–33, 97, 129–31, 187–200, 258 (n. 33), 260 (n. 7); definition of, 28–29

275

Clay County, Georgia, 4, 16, 22–23, 32, 35, 38, 49, 52, 56–57, 61, 68, 71, 89, 98, 100, 115–37, 150–51, 250 (n. 16), 261 (n. 21)
Clayton County, Georgia, 97, 102, 105–6, 113
Clinch County, Georgia, 83, 132
Coffee County, Georgia, 4, 16, 22, 32, 35, 38, 48–49, 52, 56, 61, 68, 71–72, 89, 98, 100, 115–37, 139, 151, 153
Cogswell, Seddie, Jr., 18–19, 39–40, 97–98, 100, 253 (nn. 41, 43)
Coles, Harry L., Jr., 33, 136–37, 203
Connecticut, 264 (n. 8)
Coweta County, Georgia, 175–76, 250 (n. 13)
Crawford County, Georgia, 131
Curti, Merle, 38–39

Dade County, Georgia, 75, 259 (n. 2)
Decatur County, Georgia, 82, 132
Dougherty County, Georgia, 80, 173
"Dual economy" model, 5, 181–83

Early County, Georgia, 80–81, 259 (n. 4)
Easterlin, Richard A., 100
Echols County, Georgia, 4, 16, 22–24, 31–32, 35, 38, 47, 49, 52, 57, 61, 68, 70–71, 83, 89, 98–103, 105–6, 111, 113, 115–37, 151, 153
Effingham County, Georgia, 97
Elbert County, Georgia, 108, 134–35
Emanuel County, Georgia, 4, 16, 22, 31–32, 35, 38, 49, 52–53, 56–57, 61, 63, 68, 71–72, 89, 98, 100, 115–37, 151–52, 261 (n. 21)
Engerman, Stanley L., 67

Fannin County, Georgia, 77, 106–7, 131, 156, 263 (n. 2)
Farmers, first year, 18–20, 39–40, 97–101, 105, 112–13, 124–26, 250 (nn. 12, 13), 253 (n. 43), 255 (n. 13)
"Farmers without farms," 24–26, 34, 38–44, 72, 96–107, 112–13, 121, 124–26, 201–2, 251 (n. 20), 253 (n. 46)

Fertility, influence of land availability on, 5–6
Floyd County, Georgia, 131, 134–37
Fogel, Robert William, 67
Forsyth County, Georgia, 134–35
Franklin County, Georgia, 134–37
Fulton County, Georgia, 77, 160

Gates, Paul W., 262 (n. 23)
Genovese, Eugene D., 5
Georgia, counties in. *See names of individual counties*
Georgia, regions of: Blue Ridge, 3, 5, 56, 58, 71, 74–78, 88–89, 91–92, 115–37, 147–77, 183; Central Cotton Belt, 3, 5, 56–58, 61, 71, 75, 79–81, 85, 88–89, 115–37, 144–45, 147–77, 250 (n. 13); Oak, Hickory, and Pine Uplands, 3, 5, 56–57, 72, 79, 82–83, 89, 115–37, 147–77; Piedmont, 3–5, 42–43, 48, 53–58, 60–62, 70–72, 74, 77–79, 88–89, 91–92, 102, 110, 113, 115–37, 144–77, 250 (n. 13); Ridge and Valley, 3, 5, 42–43, 58, 74–77, 91–92, 131, 148, 160, 171, 176; Wiregrass or Pine Barrens, 3–5, 53, 56, 71–72, 74–75, 78, 81–82, 89, 115–37, 146–77, 183; Long Leaf Pine Flats and Savannas, 5, 57, 75, 83–84, 89, 115–37, 147–77; Limesink, 56, 62, 72, 75, 82–83, 85, 89, 115–37, 147–77; Savannas and Palmetto Flats, 58–60, 70–71, 84–85, 134–36, 148, 163, 176; Sand Hills, 61, 75, 79–80, 89, 121, 131, 138, 175
Gilmer County, Georgia, 106–7, 131, 156, 263 (n. 2)
Glascock County, Georgia, 61, 131
Glynn County, Georgia, 84, 134–35
Gordon County, Georgia, 134–35
Gray, Lewis Cecil, 7, 95, 266 (n. 3)
Greene County, Georgia, 4, 16, 22, 32, 35, 38, 48–49, 51–52, 56, 61, 68, 70–72, 89, 98, 100–101, 115–37, 150–51, 261 (n. 21)
Gwinnett County, Georgia, 4, 16, 22, 32, 35, 38, 49, 52, 56, 61, 68, 71, 88–89, 98, 100, 102, 115–37, 146, 151–52,

160, 173–75, 251 (nn. 15–17, 22), 258 (n. 37), 261 (n. 21)

Habersham County, Georgia, 4, 16, 18, 22, 32, 35, 38, 49, 52, 56, 61, 68, 71, 89, 98–101, 106, 115–37, 150–53, 170, 251 (n. 16), 262 (n. 27)
Hahn, Steven, 1–2, 5, 42–43, 90, 93, 110, 254 (n. 52), 267 (n. 14)
Hall County, Georgia, 134–35
Hancock County, Georgia, 61, 253 (n. 37)
Haralson County, Georgia, 150
Harris County, Georgia, 4, 16, 22, 32, 35, 38, 49, 52–57, 61, 68, 71, 89, 98–101, 106, 115–44, 150–51, 262 (n. 27)
Heard County, Georgia, 134–35
Henry County, Georgia, 134–35
Houston County, Georgia, 4, 16, 18–19, 22, 31–32, 35, 38, 49, 51–52, 56, 61, 68, 71, 80, 89, 98–101, 106, 115–37, 144, 146, 148, 150–51, 262 (n. 27)
Huffman, Frank Jackson, Jr., 41–42, 254 (n. 49)

Illinois, 39, 262 (n. 23)
Iowa, 39–40

Jackson County, Georgia, 42–43, 93, 254 (n. 52), 262 (n. 33)
Jasper County, Georgia, 108
Jones County, Georgia, 61

Kansas, 264 (n. 8)

Laborers, farm and other, 13–14, 43, 46–50, 96–113, 124–26, 129–31, 153–56, 201–2, 249 (n. 6), 253 (n. 37), 255 (nn. 7, 10, 12)
Lamar, John B., 51
Lee County, Georgia, 173
Liberty County, Georgia, 84, 175
Long method, 13–26, 31–32, 126–32, 254 (n. 2). *See also* Types a to c, Types 1 to 7
Loughridge, R. H., 167–70
Louisiana, 136–37, 203

Lowe, Richard G., 40–41, 253 (n. 46), 253–54 (n. 47)
Lowndes County, Georgia, 134–37

McIntosh County, Georgia, 84
Macon County, Georgia, 80
Madison County, Georgia, 48, 108–10
Mangum, Charles S., Jr., 91, 104–5
Marion County, Georgia, 79
Maryland, 264 (n. 8)
Mendenhall, Marjorie Stratford, 7
Meriwether County, Georgia, 54–56
Michigan, 264 (n. 8)
Militia districts, analysis of, 137–44
Minnesota, 264 (n. 8)
Mississippi, 35–37, 136–37, 203
Mitchell County, Georgia, 259 (n. 4)
Monroe County, Georgia, 97
Montgomery County, Georgia, 134–37
Muscogee County, Georgia, 160

North Carolina, 93–95, 147

Oglethorpe County, Georgia, 267 (n. 10)
Overseers. *See* Census, manuscript (1860): agent, manager, or overseer
Owsley, Frank L., 2–4, 33–38, 132–37, 181–82, 185, 203, 252 (n. 30), 257 (n. 26), 266 (n. 3)
Owsley, Harriet, 33

Parker-Gallman Sample of Southern Farms 1860, 6–7
Paulding County, Georgia, 48, 101, 108–10
Pierce County, Georgia, 83
Population distributions, 156–65, 176–77
Putnam County, Georgia, 61

Railroads, 77, 167–71
Ransom, Roger L., 2, 7–8, 58, 73, 75, 81, 83, 85, 88–89, 147, 173, 175–76, 257 (n. 26), 259 (n. 4), 260 (n. 7), 266 (n. 4)
Reid, Joseph D., Jr., 2, 7, 93, 257 (n. 26)
Richmond County, Georgia, 160
Rothstein, Morton, 5

278 Index

Sharecropping. *See* Tenancy, definitions and forms of
Short method, 13, 26–33, 126–32, 187–200, 251 (nn. 24, 26), 252 (nn. 27, 29), 254 (n. 1), 263 (nn. 37, 38). *See also* Classes 1 to 5
South Carolina, 147
Spalding County, Georgia, 19–20, 250 (n. 13), 261 (n. 20)
Sutch, Richard, 2, 7–8, 58, 73, 75, 81, 83, 85, 88–89, 147, 173, 175–76, 257 (n. 26), 259 (n. 4), 260 (n. 7), 266 (n. 4)

Talbot County, Georgia, 54–56
Tattnall County, Georgia, 89, 134–37, 175–76
Taylor County, Georgia, 79, 175–76
Telfair County, Georgia, 97
Tenancy: spatial distribution of, 4–5, 8; four levels of estimation of, 4–5, 9, 34, 37, 112–37, 180–81, 201–2; definitions and forms of, 7–8, 9, 23, 90–113, 124–26, 147–48, 153–56, 183–84, 258 (n. 35); enumerator conventions for, 21–33, 62–67, 95–96, 103–12, 124–26, 153–56, 178–80, 187–96, 250 (n. 14), 251 (nn. 17, 19, 22), 252 (nn. 28–29); farm size of tenants, 144–46; cotton production of tenants, 146; wealth of tenants, 146; distributions in 1880 of, 148–77, 182, 184–85; distribution by race in 1880 of, 173–77, 185. *See also* Classes 1 to 5, Long method, Short method, Types a to c, Types 1 to 7
Tennessee, 136–37, 203
Terrell County, Georgia, 175, 260 (n. 7)
Texas, 40–41
Thomas County, Georgia, 4, 16, 22, 31–32, 35, 38, 49, 52, 57, 61, 68, 71–72, 82, 89, 98–100, 105, 115–37, 150–52, 167, 175–76, 261 (n. 21)
Towns County, Georgia, 4, 16, 22, 32, 35, 38, 48–49, 52, 56, 61, 64–66, 68, 71, 77, 89, 98, 100, 106–7, 115–37, 144, 151, 153, 156
Troup County, Georgia, 54–56

Twiggs County, Georgia, 4, 16, 18, 22, 32, 35, 38, 49, 52, 56, 61, 65–66, 68, 71, 89, 98–101, 106, 115–44, 150–51, 175, 262 (n. 27)
Types a to c (long method), 15–21, 25–26, 28, 36–44, 96–99, 132–36, 179, 181, 250 (nn. 8–9, 12), 251 (n. 23), 258 (n. 39), 263 (n. 36), 264 (n. 8); definition of, 15
Types 1 to 7 (long method), 20–27, 29, 31, 34–44, 66, 96–113, 129–36, 153–56, 179, 251 (n. 19), 255–56 (n. 17); definition of, 15

Virginia, 147

Walker County, Georgia, 131
Walton County, Georgia, 48, 101, 108–10
Ware County, Georgia, 83, 132
Washington County, Georgia, 61, 80
Wayne County, Georgia, 134–37
Wealth: North-South inequalities in, 6–7, 264 (n. 8)
Weaver, Herbert, 33–37, 136–37, 203, 252–53 (n. 35)
Westward movement, 5
Whitfield County, Georgia, 97
Wiener, Jonathan M., 2, 266 (n. 2)
Wilkes County, Georgia, 4, 16, 22, 31–32, 35, 38, 48–49, 52, 56, 61–62, 68, 70–72, 89, 98, 100, 115–37, 144–46, 150–51, 258 (n. 38), 261 (n. 21), 262 (n. 24)
Wilkinson County, Georgia, 61
Winters, Donald L., 39–40, 253 (n. 41)
Wisconsin, 38–39
Woodman, Harold D., 1–2, 7, 90, 262 (n. 23)
Worth County, Georgia, 4, 16, 22, 24, 32, 35, 38, 49, 52, 56, 61–62, 68, 71–72, 82, 88–89, 98, 100, 106, 115–37, 150–52, 167, 170, 175–76, 261 (n. 21), 263 (n. 41)
Wright, Gavin, 2, 9, 72

Yang, Donghyu, 6–7, 175, 264 (n. 8)